P9-CJY-956

Molds, Mushrooms, and Mycotoxins

Clyde M. Christensen is Regents' Professor emeritus of plant pathology at the University of Minnesota. He is the author of several other books published by the University of Minnesota Press (see

Molds, Mushrooms, and Mycotoxins

by

Clyde M. Christensen

Regents' Professor of Plant Pathology,
University of Minnesota

University of Minnesota Press,
Minneapolis

 Printed in the United States
of America at the North Central Publishing
Company, St. Paul. Published in Canada by Burns & MacEachern
Limited, Don Mills, Ontario

Library of Congress Catalog Card Number: 74-21808

ISBN 0-8166-0743-5

Two illustrations in this volume are reproduced by permission of the following: From M. J. Fiese, *Coccidioidomycosis*, 1958, courtesy of Charles C Thomas, Publisher, Springfield, Illinois; from Norman F. Conant et al., *Manual of Clinical Mycology*, 3rd ed., 1971, courtesy of Dr. Conant and W. B. Saunders Company, Philadelphia, Pennsylvania. Permission to use material from *Mushrooms*, 1954, by Albert Pilat and Otto Usak was granted by ARTIA publishers, Prague, Czechoslovakia.

Contents

List of Illustrations

Color Plates, between pages 120 and 121

Molds, Mushrooms, and Mycotoxins

Introduction

For somewhat more than 40 years much of my time has been devoted to the study of fungi themselves and to problems with which fungi are associated in one way or another. *The Molds and Man*, published in 1951, was intended to introduce the reader to the fungi in general, to their place in the general scheme of things, and to some of the strange and wonderful adaptations, dodges, and devices they have developed to enable them to compete as successfully as they do in the struggle for existence. *Grain Storage: The Role of Fungi in Quality Loss*, written in collaboration with Henry H. Kaufmann and published in 1968, summarized some 20 years of research on the relation of fungi to the deterioration of stored grains and seeds, a problem of sizable economic importance throughout the world, and one that seems likely to increase in importance with increasing world population as time goes on. The book at hand takes up a number of other fungus-related problems in some depth, but in a form readily understandable to the general reader. The subjects included were chosen, out of the many available, primarily because I have had firsthand research or teaching experience with each of them and so presumably should be able to present them with some degree of authority. The literature on some of these topics is extensive, with thousands of research papers and more than a few books devoted to them; naturally I cannot have read and digested all of these, but for those aspects of each problem that are discussed in the book the work of others has been reviewed thoroughly enough to permit what I believe to be a fair appraisal and an honest presentation of the evidence.

The first four chapters deal with fungi that are toxic in one way or another: either the fungi are themselves toxic when consumed, as with poisonous mushrooms or ergot, or the fungi, as they grow, secrete toxic compounds that diffuse into the substrate on which they are growing and so make it toxic when eaten. Poisonous mushrooms, taken up in the first chapter, have been known and written about and talked about for well over two thousand years, and one might think that by now enough information on them would have been accumulated to be codified into simple and direct commandments such as "These shalt thou eat of in plenty and enjoy; those shalt thou shun as evil." Not so; it is not that simple, and there are many peculiar twists and tangles that still have not been unraveled, and some of them probably never will be. Whether a given mushroom is or is not poisonous is of concern to others than those who go out and deliberately collect wild mushrooms to eat; it is surprising how many toddlers in suburbia forage mushrooms off the lawn in good mushroom weather, much to the dismay and alarm of the parents who envision the child succumbing to a horrid death from toadstool poisoning. (I suppose such a thing *could* happen, but I never have personally encountered or read about a case where it *did* happen.)

Hallucinogenic mushrooms also have long been known, especially among some of the Indian tribes from Mexico down through Central America and into South America, where so many species of hallucinogenic fungi, as well as other hallucinogenic plants, are found. In recent years some of these hallucinogenic mushrooms have been played up as having almost miraculous powers, enabling those who consume them to get in tune with the infinite — although such effects hardly are evident among the tribes where these mushrooms have been most used and the usually ill-nourished and malaria-ridden members of which continue to live a precarious and relatively brief existence. One of these mushrooms, *Amanita muscaria*, by virtue of its effects on those who consume it, is even claimed to have been the moving force in the formation of several ancient and modern religions, Christianity included. Each reader can decide for himself how much of this to believe.

Ergot, the subject of chapter 2, still is a common disease of wild and cultivated grasses; and although ergotism, which in medieval times scourged the human population of northern Europe, no longer is of much concern for its direct effects on people, it very definitely is of concern as an affliction of domestic animals, cattle particularly. The alkaloids of ergot and their derivatives, because of their physiological activities (witness LSD) and their use in medicine, have been investigated about as thoroughly as any similar group of compounds, and interest in them continues to increase. The mycotoxins discussed in chapters 3 and 4 have been recognized as occasional or possible health hazards to man and to his domestic animals only since about the mid-1950s; knowledge about them is so new that it scarcely has begun to get into the textbooks of colleges of medicine and veterinary medicine, although one of them, aflatoxin, is said to be the most potent carcinogenic or cancer-causing agent known and has received a relatively tremendous amount of study in many lands, with, at latest count, close to 2000 research papers devoted to it.

The next three chapters take up some of the ways in which fungi affect us and certain of the lesser animals directly. Aside from pollen, fungus spores are by far the most important causes of respiratory allergy, and the study of production and dissemination of these spores, indoors and out, has led to the development of what amounts to a special discipline, aerobiology, some of the interesting aspects of which are described in chapter 5.

Many fungi snare and consume nematodes, the mostly microscopic worms that abound in the soil, and many of them parasitize insects; the first disease of any kind proven to be caused by a fungus was, in fact, a fatal malady of silkworms shown by Bassi, in 1835, to be caused by a fungus that later was named *Beauveria*. It once was thought that these fungus parasites of insects might offer an effective means of insect control (this turned out to be not so) and a good deal of work was devoted to exploring this possibility, much of it before 1900. Some of these predatory and parasitic fungi also live rather unusual and almost implausible lives, and from that standpoint alone they are worth getting acquainted with; they are described in chapters 5 and 6. Chapter 7 is devoted to the superficial and deep-seated fungus infections or mycoses of man and of some of his domestic animals; several of these are pretty horrid, but as killers they cannot compare with the more virulent bacterial, viral, and protozoan parasites. They are much more common than generally is realized, however, and at times they constitute a fairly serious public health problem.

Chapter 8 deals with the nature, cause, and prevention of wood decay in trees and buildings, a subject about which a great amount of information has been accumulated over the past century, since the first book on the subject was published, but about which people in general seem to know very little, and most of that little probably is not correct.

The last chapter takes up some aspects of organic evolution in general, as a background for presenting the few theories and still fewer facts on the evolution of fungi, and gives a brief summary of some of the ways in which fungi enter into our lives and economy now, and a guess about what they might amount to in the future. It seems probable that the fungi, along with many other of the less conspicuous but better adapted organisms on the earth, will long outlast us; perhaps by studying the ecology of these organisms we may eventually learn to live in harmony with one another and with our fellow beings, although this seems highly unlikely. Nature operates mainly by struggle and competition, not by forebearance and compassion, and man is no exception to this general rule. If this seems a hard conclusion, to a biologist — or at least to this biologist — it is a reasonable one. This fragile earth ship may eventually fail, as portions of it already have begun to do. Perhaps in time we will have learned enough to preserve and protect it — only time will tell.

Sufficient literature has been cited for each chapter to permit those interested in the subject to continue further reading on their own. The references are gathered at the end of the volume.

1

Poisonous and Hallucinogenic Mushrooms

Before we get to the central topic of this chapter, poisonous and hallucinogenic mushrooms, a few words on the general subject of picking and eating wild mushrooms. Wild mushroom hunters, or — grammatically at least — more accurately hunters of wild mushrooms (a recent magazine article, describing the death of people who collected and ate wild mushrooms in California, referred to them as "fungus freaks," which seems unkind), are increasing, and so, doubtless, cases of illness or death from eating the poisonous kinds also will increase. Remarkably enough, many of those who collect and eat wild mushrooms, year after year, have only a very superficial acquaintance with any of the technical aspects of mushroom identification — and, seemingly, they couldn't care less. Some do not even bother with books or manuals that would aid them in acquiring at least a working knowledge of the subject; they simply go out and collect wild mushrooms and eat them, and usually they get by quite all right. Or they know, or they think they know, some of the poisonous species of *Amanita*, and avoid those, but gather and eat almost anything else, a practice not at all approved of by the conservative and stodgy pros, who eschew that adventurous life style and continue to insist that the only safe procedure is to know with absolute certainty the identity of every specimen picked, and to pick only those that are known to be safe. Not even that approach can be guaranteed to keep an eater of wild mushrooms out of trouble, since, with wild mushrooms as with many other edibles, what is one man's meat may be another's poison. Even the succulent and savory morel, certainly one of the choicest of the edible wild mushrooms, has been known to cause illness when eaten — illness of relatively brief duration, to be sure, and of not too serious a nature, but illness nevertheless. Some kinds of perfectly edible and delicious wild mushrooms may be infested with the larvae of fungus flies and with the bacteria, nematodes, and other small forms of life that accompany such infestation, and they may be unwholesome on that account, as indicated in table 1-1.

Naturally, people differ in their opinions about the edibility of wild mushrooms; on the one extreme are those who think that none of the wild mushrooms are edible and who, when they want mushrooms, buy the cultivated kinds in the store. My own personal opinion is that very few of the wild mushrooms equal the cultivated *Agaricus campestris* (or *Agaricus bisporiger* if you prefer) in flavor, and that many of the wild kinds are flat and insipid in flavor or unattractive in texture. Obviously this view is by no means shared by many of my fellow hunters of wild mushrooms, some of whom maintain with all sincerity that many kinds of wild mushrooms are far superior in flavor and texture to any kind of cultivated mushroom. Some wild mushrooms, such as the Foolproof Four described in *Common Edible Mushrooms* (4 — see the list of references beginning on page 249) or those illustrated in the folder on *Edible Wild Mushrooms* (5), probably would rate near the top of anyone's list of choice edible mushrooms, but to other mushroom buffs other kinds are equally good.

And so to the poisonous mushrooms.

Poisonous Mushrooms

If people who have had little or no formal or even informal training in the identification of mushrooms can just go out and collect wild mushrooms more or less at random, and eat them, and can do this year after year without ever coming to grief, the really poisonous kinds must be relatively rare. With a few exceptions, they are, or they are found only briefly and at special times and places. Probably no more than 50 of the 5000 or so species of mushrooms are poisonous, and of these 50 poisonous kinds, only a few are lethal. So far as I am aware, no medical statistics — no ''morbidity and mortality'' tables, as they are called — dealing with poisonings from mushrooms are gathered in the United States, but table 1-1, from Pilat (13), gives such statistics for 1943 and 1944 in Switzerland. Note that two fatalities were attributed to *Amanita phalloides* and two to *Inocybe patouillardi* (which does not occur in the United States, although many other speices of *Inocybe* do, some of them or perhaps most of them poisonous, but none of them lethal, or at least none of them known to have caused fatal poisoning, for reasons to be given shortly). Note also that 74 of the 356 people who became ill after eating wild mushrooms did so after consuming good edible species that were partly spoiled. Unless the Swiss fungus hunters are far sharper than those in most other countries, there is always the possibility that an occasional misidentified and toxic specimen lurked among the ''good edible species.''

Table 1-1. Mushroom Poisonings in Switzerland in 1943 and 1944

Species	Number of Incidents	Number of Persons Ill	Number of Deaths
Amanita phalloides	4	10	2
Amanita pantherina	10	36	0
Tricholoma pardinum	21	141	0
Clitocybe rivulosa	1	2	0
Clitocybe nebularis	4	11	0
Inocybe patouillardi	2	6	2
Entoloma lividum	12	39	0
Acrid *Russulae*	2	6	0
Coprinus atramentarius	1	1	0
Agaricus xanthodermus	1	4	0
Boletus satanas	1	4	0
Ramaria pallida	4	17	0
Digestive troubles, almost all caused by good edible species which were eaten when spoiled	18	74	0
Poisonings with species not determined	12	35	0
Total	93	356	4

Source: Adapted from Pilat (13).

Pilat divides the toxic or poisonous mushrooms into seven groups, according to the toxin or toxins that they contain and according to the mode of action or effects of the toxin. Simons (18) follows about the same pattern. One of the toxins produced by several species of mushrooms is muscarin, so named because it first was found in *Amanita muscaria* (to be taken up in the section dealing with hallucinogenic mushrooms); it also is found in one or more species of *Clitocybe* and in many species of *Inocybe*. Kauffman (10) describes 52 species of *Inocybe*, and Pilat quotes an investigator of mushroom toxins to the effect that all 33 species that she tested contained some muscarin and therefore could be considered more or less toxic. Some of the 52 species of *Inocybe* described by Kauffman are fairly widespread in the United States — a few specimens of *Inocybe rimosa* come up in my lawn every summer — and yet there are no more than a few scattered cases of mild poisoning attributed to consumption of *Inocybe*. If there are so many species of *Inocybe*, and if nearly all of them contain some toxin, and some of them contain moderate to large amounts of toxin, why are not more people poisoned by eating them? The answer is simple: few specimens of *Inocybe* are eaten. Kauffman characterizes the genus as "absolutely negligible and uninviting as food," with which I agree, and evidently so do most people who gather wild mushrooms for eating. Most species of *Inocybe* are small and fragile; a few of the species sometimes appear in moderate abundance in a very limited area and for a brief time, but most of them occur as a few scattered specimens here and there. Of those few species that do occur in some numbers briefly, it would be necessary to collect several hundred specimens to get enough, when boiled or broiled or fried, to cover a piece of toast, and even that small portion would consist of an unattractive semiliquid brown glop or gunk. If it is mushroom weather and the species of *Inocybe* are numerous, a lot of larger and more attractive mushrooms will be growing too, and few people are likely to deliberately gather *Inocybe* for the table. As support for this, in a good mushroom year, it is not unusual for people to bring in to my laboratory at the University of Minnesota baskets and bucketsful of wild mushrooms to be identified; over the several decades that this has been going on, I do not think that I have seen so much as a single specimen of *Inocybe* in their collections.

Approximately the same combination of characteristics applies to some of the other species of potentially toxic wild mushrooms, and this goes a long way toward explaining why most cases of serious or fatal poisoning resulting from consumption of wild mushrooms involve only a relatively few species. The more important of these will be taken up in the following discussion.

The *Amanita Phalloides–Amanita Verna* Group

Simons says that the *Amanita phalloides* group is estimated to be responsible for 95 percent of the fatal cases of mushroom poisoning throughout the world. He mentions one outbreak of mass poisoning that occurred near Poznan, Poland, in 1918, in which 31 school children died from eating a dish of mushrooms containing this fungus — presumably in their school lunch, one that must have set an all-time low for school lunches. Detailed accounts of some cases of poisoning by *Amanita phalloides–Amanita verna* will be given below.

Kauffman describes 22 species of *Amanita*, of which eight are listed as "deadly" and seven as "suspected," two "probably poisonous," and the remaining few as "edible but use caution." I do not know exactly what this "caution" involves, but presumably it is caution in identifying, since once the mushroom is in the pot and on the plate, if it is poisonous it does not matter much how cautiously you partake of it. In my opinion, anyone who would eat *any* species of *Amanita*, or even any mushroom that possibly could be confused with *Amanita*, has a somewhat casual attitude about life, akin to that of skydivers and of those who run snowmobiles over thinly frozen lakes. Kauffman says, "But one who has not a thorough knowledge of most of the mushrooms" (a very tall order, indeed), "including their microscopic characters, would be unwise to eat any of the species of *Amanita*, since the poisonous species sometimes approach the edible ones quite closely in general appearance. And to serve them to others under ordinary circumstances is worse than criminal." I would amend that to read "under any circumstances."

The genus *Amanita* is characterized by (1) white gills and spores; (2) free gills — that is, they come up close to the stem but do not touch it; (3) a veil in young specimens that extends from the edge of the cap to the stem — as the cap expands, this veil breaks at the edge of the cap and remains as a ring or miniskirt on the stem; (4) a volva or cup surrounding and enclosing the base of the stem. Often the veil collapses on the stem and becomes practically invisible, and often the volva or cup is buried in the soil and so is unseen unless its presence is suspected and the base of the stem is dug out. Some species do not have a definite cup, but only an enlarged base of the stem, with maybe some ridges on it that suggest the remains of a cup.

Of the highly poisonous *Amanita phalloides–Amanita verna* complex, *A. phalloides* evidently is more common in Europe, and *A. verna* and its varieties are more common in the United States, although both occur on both continents. As a general rule, neither is likely to be abundant, but in a given wooded area at a given time, usually about the middle of August, one may find dozens of specimens of *A. verna*. It is listed as occurring in forests, rarely in fields and grassy places, but I never have seen it outside of the forest, and certainly never in lawns. It has been *reported* in lawns, but these reports never have been confirmed, so far as I know, by anyone who was really competent to identify the fungus. Figure 1-1 illustrates *A. verna*.

Examples of poisoning by Amanita. Simons says, "Mute testimony to the terrible suffering of *A. phalloides* victims is provided by seven mummies, a single family, now in the crypt of the Tour St. Michel, Bordeau, France. Workmen moving these mummies, along with 63 others, from a fifteenth century cemetery to the crypt in the year 1810 were so disturbed by the expression of pain, still clearly evident on the faces, that special efforts were required by the priests and civil authorities to keep the work in progress."

One might well ask how it was known that these were the victims of *Amanita phalloides* poisoning. The facial expression of a 400-year-old mummy is diagnostic? If it was poisoning at all, then why not by ergot, since ergotism was rife at the time and place where these people lived and died, as will be seen in the next chapter. And the mummies retained, for 400 years, the grimaces of death? No 400-year-old mummy — or 400-year-old mummie either, for that matter — is a thing of beauty; not even a French one. I question this account.

Granted that the symptoms of poisoning by *Amanita phalloides* or *A. verna* and their close relatives are horrid enough. Two cases, from books on mushrooms, will suffice.

Figure 1-1. *Amanita verna*, the extremely toxic fungus responsible for most cases of fatal mushroom poisoning

From Fischer (6) comes this account:

In September, 1911, six persons were poisoned, two fatally, in Cleveland. The children, aged four and six, had a little gravy and recovered after nausea, vomiting and diarrhea. Mr. C., aged 67, ate some at supper, felt bad during the night but ate more for breakfast! About noon violent illness began with intense pain in the epigastrium, vomiting and diarrhea with loss of control, clonic spasms and great prostration. Urinary suppression was obstinate and lasted till death, three days after the first meal. Mrs. C., aged 65, ate one forkful at the first meal but did not like the taste. Profuse vomiting and diarrhea with great prostration began one hour after Mr. C.'s symptoms. She recovered rapidly after two days. The daughter-in-law, aged 40, ate at the second meal and, though feeling hot and feverish, ate more at noon! Eight hours later she had exactly the same symptoms as Mr. C. The physician arrived early on the next day. He gave oil, began stimulations with strychnia, nitroglycerin, aromatic spirits of ammonia and, after removal to the hospital, saline solution continuously and oxygen. Some improvement was noted except that the heart action was weak and intermittent and the extremities could not be kept warm. Hiccough for two days, great agony and unconsciousness preceded death on the seventh day. A son, aged 19, ate the second meal — breakfast. Though feeling bad, he worked until 4 P.M. By 9:30 he presented the same symptoms as mother and grand-father. After ten days of apparently as grave illness as theirs under ''most terrific stimulation'' (nitroglycerin, strychnia, oxygen and salines) he was reported out of danger though ''looking like a corpse.''

Güssow and Odell (8) cite this case:

In the early forenoon of Saturday, September 11, 1926, M., a day labourer, aged fifty-five, collected in a thin woodland near London, Ont., three or four quarts of mixed white mushrooms. They were cleaned and stewed in milk with chopped parsley and onions. The parents with two children, John, aged twelve, and Annie, aged seven, had dinner about eleven o'clock and ate a portion of the mushroom stew.

In the mid-afternoon a Polish acquaintance, Stan. S., aged forty-five, called at M.'s home. He learned of the mushroom delicacy they had had for dinner, regretting that he had not gone mushroom-collecting with M., and was pleased when he was given a bowl of the stew, which he took away to the home of Joseph S. Between three and four o'clock these two men sat down to enjoy the treat. Jos. S. sampled two or three spoonfuls, declared they were bitter and that he did not like them and stopped with that. Stan. S. said they were good, that he liked them, and finished his dish. The former assured me himself that he did not swallow more than a good teaspoonful. By 5.00 P.M. Mrs. M. began to feel pain in the region of the stomach, and nausea, succeeded by dizziness and a feeling she described ''as if I were drunk.'' Thereupon she dosed herself heavily with castor oil. Shortly afterwards her husband came into the house feeling ''queer.'' Then Stan. S. called to thank them for the mushrooms, and finding two of them ill assured them that it couldn't be the fault of the mushrooms seeing that he was feeling fine. In the late evening the twelve-year-old boy alarmed the neighbors with the story that his father and mother were dying. By nine o'clock the two children as well as the parents were very ill. Medical aid arrived. Cramps, particularly in the legs, vomiting, diarrhea, and other evidences indicated amanitine poisoning to the physicians, and they did the best in the case that medical science knows. The family, including a nursing infant, were removed to the hospital. After the lapse of a like number of hours Stan. S. became similarly ill and was taken to the hospital. Jos. S.'s turn came next, but he remained in his home and was nursed by his family.

The hospital records show, of course, the progress of the poisoning in each case, and the remedial measures and medicaments employed to meet the varying conditions as they developed. Taken altogether they exhibit variation and repetition of the following effects: cramping pains more or less violent, often in the limbs; vomiting, sometimes of greenish liquids; diarrhoea, greenish liquid stools, passing of blood; fierce thirst; local or general severe soreness of the muscles; very rapid, weak, thready pulsation; bluish or greenish jaundice; alternation of drowsiness and delirium; rigidity of the limbs; brief coma. In the tale of medication employed, when and as needed, were stomach-washing, colon-flushing, morphia, spiritus frumenti, atropin, digitalis, hyoscine, adrenalin.

Annie M., the seven-year-old child, was the first to be relieved by death. In twenty-two hours from the eating of the mushrooms her eyes were becoming starey, her extremities rigid, and her throat unable to swallow. Before the twenty-fourth hour had elapsed her consciousness had ceased; adrenalin failed to whip up the heart to further action. She was dead.

The father, a man of rugged, muscular frame, was next to go. On Monday morning, in his delirium, he got out of bed and as late as four o'clock in the afternoon he was struggling to rise. At 4:20, in the forty-first hour after the fatal meal, he had ceased to breathe.

Stan. S., an able-bodied and younger man than M., whose mushrooms were eaten four or five hours later, survived him by twelve hours, being a period of two days and fourteen hours from the time he supped the dish. He was delirious and talking wildly when his eyes began to glaze, his limbs stiffened, his face became gray, and his pulse stopped.

Jos. S. who stated to me that he hardly more than well tasted the stew, kept his bed at his home for five days. His acute attacks would seem from the accounts I obtained to have been quite severe, but yet less violent and less frequent than those already referred to. On the fifth day, he had one or two of these attacks, but at four o'clock in the afternoon his family thought he was past the crisis and were hopeful of his recovery. And yet, in less than two hours, with his physician at his bedside fighting for his life, his heart failed.

John M., the boy, suffered experiences similar to those of the others. At eight A.M. on the fourth day he stiffened out, and with head thrown backwards, and eyes starey, he mumbled deliriously through an attack that lasted about five minutes. Revived by stimulants, he vomited some greenish fluid and from that time began to improve without relapse. He left the hospital on the twenty-fifth day, still paralyzed in his legs. With the help of crutches he began to use his legs, gradually recovered their tone, and now at the end of two and a half months is nearly well.

The infant that fed at the breast of the poisoned mother on the first half day, showed no toxic effects.

The mother herself, who thinks she ate as much of the stew as any of the others, suffered terrible pains in all her muscles and passed considerable blood, but she seemed to escape the intense severity of the onsets endured by those who died. She may have responded more favourably to the medication, or the heavy dosing with castor-oil and the early vomiting may have had beneficial effects. She and her son left the hospital the same day. Now in the eleventh week she can still feel the effects of the poisoning in the muscles of her arms and calves.

Dr. John Dearness, who contributed the account above, was a professional mycologist in Canada. He identified the mushrooms, or the remains of them, that were in the fatal stew, and found them to include *Clitocybe*, *Lactarius* or *Russula*, and *Amanita*, probably *A. verna*. I've sampled a few species of *Clitocybe*, *Lactarius*, and *Russula*, but never have found them to be very attractive for eating; besides, most caps of *Lactarius* and just about all those of *Russula* are pretty well infested with the larvae of fungus flies by the time the caps come above the ground. So for a mess of wild mushrooms made up principally of kinds that at best are by no means choice, plus some specimens of *A. verna*, six people experienced the horrible suffering described above, four of them died, and a couple of months or more later the two survivors still had not totally recovered. M., the man who gathered the mushrooms that brought about the tragedy, probably had been gathering and eating wild mushrooms all of his adult life and probably thought that he knew the difference between edible and poisonous mushrooms. Until this time, he just had not happened to come across any really poisonous kinds, and so was protected from harm by the relative scarcity of these mushrooms rather than by any knowledge of his own.

Note that in the account of this case in Canada no symptoms of discomfort appeared until some hours after the mushrooms were eaten. In the cases described by Fischer, two of the people, although feeling slightly unwell between the time of first eating the mushrooms and the next mealtime, still had another helping of the lethal mushrooms at the second meal.

The toxins. There are two groups of toxins that are present in these mushrooms — the phallotoxins, composed of phallin and phalloidin, and the amanitins, made up of alpha amanitin and beta amanitin. According to Floersheim (7), the gastrointestinal symptoms, which appear first and some hours after the mushrooms are consumed, are produced by the phallotoxins. The amanitins are 10 to 20 times as toxic as the phallotoxins, and cause the hepatorenal, or liver and kidney, damage that results in the symptoms appearing later and that in many cases ends in death. Simons quotes figures from other investigators who determined the amount of amanitins in *Amanita phalloides*, *A. verna*, and *A. bisporiger* (the latter a variety supposedly distinguished by having two spores per basidium, whereas *A. verna* has four; I have had specimens in my laboratory in which two-spored and four-spored basidia were present in about equal numbers, and often alternating with one another on the same gill, so I question this distinction). The amanitins were present in amounts of 0–1.7 mg/gram of dried mushroom tissue in *A. verna*, 2.25–5.0 mg/gram in *A. bisporiger*, and 1.9–2.43 mg/gram in *A. phalloides*. Only one to ten specimens of each species were tested, so this does not mean too much except that these toxins are present in small but variable amounts. One milligram of toxin per gram of dried tissue (1 mg/g) is one part by weight of toxin to a thousand parts per weight of the dried mushroom. In toxicology it is customary to speak of parts per million (ppm), or parts per billion (ppb). Thus 5.0 mg/g, the highest concentration of amanitins reported, equals 5000 ppm of toxin, which is a fairly high concentration of this compound, as fungus toxins go.

How much is required to poison a person fatally? This varies some, of course, but we can work out the lethal dose in a general way. One gram of dried mushroom is equal to about 30 grams of fresh mushroom, or one ounce — about a good tablespoonful. From the case histories given above it is evident that at times a tablespoonful of mushrooms of which *Amanita verna* or *A. phalloides* made up only a part was enough to cause serious illness or death. Jos. S. assured Dr. John Dearness that he did not swallow more than "a good teaspoonful." The man was a laborer, so he probably weighed at least 75 kilos (165 pounds), which is equal to 75,000 grams or 75,000,000 milligrams. Even if the *Amanitas* in the mess of which he partook were unusually potent, he could not have consumed more than 5–10 mg of amanitin, so that one part of amanitin per 10,000,000 parts by weight of Jos. S. was enough. This agrees approximately with Simons who says that, on the average, specimens of *A. phalloides* in Europe contain about 10 mg of phalloidin, 8 mg of alpha amanitin, and 5 mg of beta amanitin per 100 grams of fresh mushroom tissue. One hundred grams of fresh mushroom tissue is equal to about three grams of dried mushroom tissue. He says that the lethal dose of amanitin for humans is about 0.1 mg per kg of body weight, or 1 part in 10,000,000, which equals 100 ppb.

In a case of poisoning by *Amanita verna* in 1972, in Minneapolis, Minnesota, a husband and wife consumed sautéed *A. verna* (I later identified the remains of the mushrooms) in a salad Sunday evening. He ate more than she did (maybe to prove to her, or to himself, that the mushrooms were good, since he had picked them). Sixteen hours later they developed the typical gastrointestinal symptoms — nausea, vomiting, diarrhea. They entered the hospital Tuesday. Both were treated symptomatically, with intravenous fluid replacement. The wife's gastrointestinal symptoms persisted for three days, his for six. The patients were discharged after four and six days respectively, with no evidence of any liver or kidney damage. The physician who treated them supposed that the mushrooms they ate contained phalloidin toxins but not enough amanitin toxins to cause detectable harm — a reasonable assumption. They were lucky.

Treatment. According to Pilat rabbits are able to eat "a considerable quantity of *A. phalloides* without harm." (I've never seen a fruit body of either *Amanita phalloides* or *A. verna* so much as bitten into by squirrels or rabbits; in fact I never have known a rabbit to eat *any* kind of mushroom, although red squirrels commonly stash away dried fruit bodies of *Russula*, or hide fresh fruit bodies to dry; maybe they have gourmet rabbits in France.) In any case, this real or supposed resistance of rabbits to poisoning from eating these toxic fungi led a French professor to suppose that the stomach of a rabbit contained or produced some "antisubstances" that modified or counteracted the effects of the poison. So for treatment of those poisoned by *A. phalloides* he devised a paste made of three (that magic number three) hashed raw stomachs and seven (that magic number seven) raw brains of rabbits, which the patient was to eat and retain. Pilat says, "As, however, the patients easily vomit, and this food in itself rouses disgust, they often can not keep it down. In other cases, however, the vomiting stops after consuming this paste." (What he means, of course, is that the vomiting stops after the patient consumes the paste, but that isn't what he says.) Maybe the vomiting would have stopped anyway. Also this horrid mush could have no effect on the later intoxication of and damage to the liver and kidneys if the poison already had got that far. I have not been able to find any explanation of why the brains of rabbits were included — maybe the French savant figured that they gave punch and verve to the concoction.

According to Ramsbottom (14), "The Institut Pasteur has produced an antiphalloidian serum by immunising horses. This has given consistently good results when injected hypodermically or intravenously shortly after the first sign of poisoning." His book was first published in 1953 and this "antiserum" must have been available for some time then. No information is given concerning its use or clinical experience with it. Chemists and toxicologists writing on *Amanita* toxins and poisonings in the last few years, in both the United States and Europe, do not mention it. If it were a really effective antidote for poisoning by *Amanita phalloides* and *A. verna*, it seems strange that it has not received wider publicity and wider acceptance.

Thioctic acid (alpha-lipoic acid) has been suggested as an effective treatment for *Amanita* poisoning; Simons says that Kubicka, in Bohemia, in 1964 reported that 39 out of 40 victims were saved by administration of an infusion of thioctic acid plus salts and dextrose.

In St. Paul, Minnesota, in 1971 three men were severely poisoned by eating *Amanita verna* (one of the men later said that the mushrooms looked so good, they must be good to eat!). Thioctic acid was administered to two of them, and they recovered. The third victim refused, on religious grounds, to receive thioctic acid treatment; he also recovered.

Floersheim says, "The fatalities occurring after ingestion of the poisonous mushroom *Amanita phalloides* are thought to be due to alpha amanitin. Cytochrome-c provided antidotal effects against a lethal dose of this toxin in mice. Significant survival was obtained even when the treatment was withheld for eight hours." So far as I am aware, no tests have been reported with treatment of humans with cytochrome-c after poisoning by *Amanita*.

Other Poisonous Mushrooms

Some authorities list *Amanita muscaria* unequivocally as deadly poisonous, and at times it is, but in recent years it has received much more notoriety for its hallucinogenic than for its lethal properties, and so it will be considered in a later section of this book along with some species of *Psilocybe* and *Stropharia* that also are hallucinogenic. A number of other species described by Pilat and by others as poisonous are responsible for relatively few cases of poisoning, and none of those fatal. Two of these species, however, *Gyromitra esculenta* and *Coprinus atramentarius*, will be discussed because of the somewhat unusual circumstances associated with their toxicity.

Gyromitra esculenta. The case for — and against — this mushroom is peculiar, to say the least. *Gyromitra* means gyrose or convoluted head, and *esculenta*, of course, means esculent, or edible, with the connotation that it is not only edible but very good. It is a common early spring mushroom, with a hollow and fragile spongy stem surmounted by a brown, convoluted cap — one of several common saddle fungi. Seaver (17), an authority on these fungi and their relatives, does not accept the name *Gyromitra* at all, but uses the genus name *Elvela*, which others spell *Helvella*. Ainsworth and Bisby's *Dictionary of the Fungi* (1) lists 25 species of Helvella, whereas the book by Seaver includes 15 species in the United States. To Seaver, what others refer to as *Gyromitra esculenta* is merely a gyrose form of another species, *Elvela infula*. Some species of *Elvela* (or Helvella) or *Gyromitra* are so characteristic that, once seen, they can be recognized easily and with certainty. Others cannot be, and this probably applies to *Gyromitra esculenta*, since we already have Seaver considering it to be just a form of another species. *Helvella underwoodii* looks a good deal like *Gyromitra* or *Helvella esculenta*, but is poisonous. Smith (19) says that some of his students who ate small bits of the fruit bodies of *Gyromitra esculenta* (they knew that it might cause symptoms of distress) were somewhat poisoned by it; in that case there could hardly be any doubt that the fungus was *Gyromitra esculenta*, because Smith does not err in identification of fungi. Most books on wild mushrooms include only one or two species of *Gyromitra*, and amateur mushroom hunters, if they depend upon the help of a book at all, will rely on the one they happen to have. Miller (12), a professional mycologist who has collected and studied fleshy fungi in many parts of the United States, says of *Gyromitra esculenta*, "This is a really variable group, with toxic and edible species resembling each other confusingly. The chance for error is great." If the species in this group are difficult for Miller to separate from one another, they must be more so for an amateur.

I have looked up the comments on *Gyromitra esculenta* in 20 books on fleshy fungi in four languages, dating back to 1847. The comments range from simply "Edible," with no qualification whatever, to "Deadly poisonous," with no qualification whatever. Güssow and Odell, whose expertness regarding fleshy fungi is above question, say that they have eaten this mushroom frequently and without any ill effect, although they were aware that it had at times caused illness and even death of those who consumed it. They conclude, "We are firmly of the opinion that *Gyromitra* is not only safe to eat, but a very desirable fungus for the table, but we emphatically warn against the use of old and even slightly deteriorated specimens." Others also attribute the occasional and sporadic toxicity of this mushroom to partial spoilage. Pilat gives several theories that have been put forward to explain its toxicity, to wit: (1) there are several closely related species, one of which is poisonous, the others edible; (2) some individuals happen to be sensitive to it; (3) it contains helvellic acid, a known toxin, but one which is destroyed by cooking; (4) the toxic specimens are partly decayed. He leans to (4) as the most probable explanation. Yet Ramsbottom, writing in 1953, says that large quantities of this fungus are gathered in Poland and exported to Germany — presumably dried. It is highly unlikely that those who gathered these large quantities would carefully examine each specimen for decay; or that they could recognize the early stages of decay; or that the gathered specimens would be handled with extreme care until they were dried or otherwise processed. He also says that all cases of poisoning from this fungus — he refers to them as "accidents" — have been reported from Germany, none from France. Pilat says that slightly decayed specimens are very difficult to detect, since what appears to be a young and fresh specimen may actually be several days old, and senescent.

I have at times put pieces of fresh cap of *Gyromitra esculenta*, or at least what I and some of my colleagues identified as *G. esculenta* — even I am losing confidence now — on agar in petri dishes; ascospores are shot out onto the agar, germinate, and grow into mycelium. The idea was to try to grow some of these fungi artificially, much like the cultivated mushroom is grown, and really clean up; we were mainly interested in morels, but tried some of their relatives too (it didn't work out; we could get mycelium, but no fruit bodies). In every case, the ascospores so discharged from what looked like fresh, crisp, and sound caps had bacteria with them. So if the presence of bacteria is evidence of incipient decay, then, according to my experience, there are no really sound specimens of *G. esculenta*, and every collection of them must include some specimens in various degrees of spoilage. Many wild mushroom hunters are not overly fussy about the condition of the specimens they eat; they collect, cook, and eat specimens of *Lactarius*, *Russula*, and other kinds that are partly tunneled through by the larvae of fungus flies and are partly decayed, and do so without harm. This is no proof that "slight decay" in *Gyromitra* is not responsible for its toxicity, but it leads one to question whether that is the answer.

Some say that the toxic principle is water soluble, and that if the specimens are parboiled and the water poured off, the toxin is gone. Krieger (11) quotes a member of the New York State Museum, and therefore supposedly reliable, who had gathered and eaten this fungus for 30 years and who also regularly consumed the tasty juice in which it was cooked, and never experienced any ill effects. I once talked about wild mushrooms to the Minneapolis Mycological Society — a group of avid mushroom hunters and mushroom eaters. When I described

Gyromitra esculenta as a species of doubtful edibility and possibly even lethal, I was practically laughed off the stage; many of them had eaten it in quantity every year and none of them had ever experienced anything but joy and wellbeing following the eating. I slunk back to my ivory tower in disgrace. I am sure that those people took no special precautions to avoid all specimens with the slightest detectable taint, which makes me doubt the theory that the poison is due to partial decay. Whatever it is due to, cautious or timid people, among whom, so far as eating wild mushrooms is concerned, I place myself in the forefront, will do well to leave *G. esculenta* in the woods.

Coprinus atramentarius. Cases of poisoning by this common, widespread, and generally good mushroom are relatively rare, but they seem to be well documented. (The mushroom is pictured in plate 6.) Briefly, some people who consume an alcoholic drink before or after eating *Coprinus atramentarius* are mildly to not-so-mildly poisoned by it — the face becomes flushed, the heartbeat goes up to 130–140 per minute, accompanied by profuse sweating, ringing in the ears, and a feeling of anxiety and unsteadiness, perhaps prostration. The symptoms continue for various lengths of time, but usually pass within a few hours, with no lingering aftereffects. If alcohol is consumed again within 24 to 48 hours, the symptoms may recur, but in lesser degree. This seemingly is straightforward enough — some people just happen to be sensitive to a combination of alcohol and *C. atramentarius*. The fungus, that is, contains a toxin that is soluble in alcohol, or contains some substance that is transformed by alcohol into a toxin — but only in some people, not all. Or do only certain strains of the fungus contain this still hypothetical toxin?

Some mycologists suggest that the mushroom contains a substance akin to "antabus," a drug sometimes administered to alcoholics which produces symptoms similar to those listed above, the idea being that the person so affected will give up alcohol, or, alternatively, will give up antabus. That the toxic substance in *Coprinus atramentarius* actually is antabus seems to me to be unlikely, because so few people have this reaction to the combination of *C. atramentarius* and alcohol, whereas a combination of antabus and alcohol invariably produces the reaction. Sometimes, but less frequently, *Coprinus micaceus*, the small inky cap that is so common and abundant that it must be eaten by literally millions of people each year, will produce the same effects with alcohol, but again only in some people, not in all.

Pilat cites eight accounts of this sort of poisoning by *Coprinus atramentarius*, by seven authors, dating back to 1906. In one case which he characterizes as "specially heavy," the patient had eaten, at one sitting, or one feeding, *six pounds* (my italics) of *C. atramentarius* and had drunk "plenty of wine" with it. This gourmet suffered, in addition to other symptoms, vomiting, stomach pains, and diarrhea. After eating six pounds of mushrooms — say six pints at least — plus an unspecified but perhaps hearty helping of potatoes, sauerkraut, and other goodies such as sauerbraten or pickled pigs' feet, all washed down by a couple of quarts of wine, no wonder the man was ill for a while, toxin or no toxin.

The relative rarity of this sort of poisoning is attested to by its lack of mention in many books dealing with wild mushrooms, including my own *Common Edible Mushrooms*, published in 1943 — at least until a limiting cautionary clause recently was added. I, and a number of other authors of mushroom fieldbooks for the layman, simply described *Coprinus atramentarius* as edible, because we and our friends and acquaintances had eaten it repeatedly (although never, I am sure, in quantities of six pounds per person) with no ill effects at all. I do not specifically know that we consumed any spiritous liquors before or after eating it, but I think it highly probable that at times we did so. Also, the fungus has been eaten since ancient times, and in many lands where most people regularly consume wine, beer, schnapps, or other alcoholic drinks along with their meals and between meals. So in the light of this it was only natural that those who had for years been collecting and eating wild mushrooms, *C. atramentarius* among them, should list it as edible without further comment. Another indication of the relative rarity of this sort of poisoning: The book *Common Edible Mushrooms* was published in 1943, as stated above, and up until 1972 had gone through several reprintings, and for some time had been one of the books recommended by the National Wildlife Federation. In that year an outraged reader wrote to the Federation accusing me of being "criminally negligent" in listing *C. atramentarius* as edible, and suggesting that a public retraction be made. So far as could be determined from the letter, she herself had not had any personal experience with poisoning by this mushroom, but she had read of a case. By that time I had read of a couple myself and in a colored folder, *Edible Wild Mushrooms,* published in 1968, a cautionary sentence was included with the description of the inky caps to the effect that some people were poisoned by a combination of this mushroom and alcohol. "Intoxicated" is the technical word for this. Some people sometimes become intoxicated from consumption of alcohol alone, but this cannot explain the sort of intoxication discussed here, because of the exceedingly small quantity of alcohol needed to set it off.

Some mycologists, in fact, suggest that *Coprinus atramentarius* is not at fault at all, but that the trouble results from the mushroom hunter collecting and eating a closely allied but different species of *Coprinus.* Smith, who has had some decades of experience with wild mushrooms and with eaters of wild mushrooms, says that he personally has encountered three people with this sensitivity to intoxication by inky cap–alcohol combination, but he does not say whether the mushrooms in question were identified by him.

I looked up what was said about the edibility of *Coprinus atramentarius* in the mushroom books available to me in the library of our Plant Pathology Department and in my own home library, a total of 26 books, in eight languages. In ten of these books, some of fairly recent date, the fungus simply is said to be edible, or edible when young and before the caps begin to liquefy. Most of the remaining ones state that it is said that consumption of the fungus, followed or preceded by alcohol, results in mild disturbances. A few list it simply as "toxic" or "inedible." The sensible thing to do, then, if you eat wild mushrooms and drink alcohol, is to try the combination and see if you are sensitive to it and, if you are, to eliminate this mushroom from your list of good kinds. As for treatment of this sort of poisoning or intoxication, Simons has only one word: "Reassurance."

In the early fall of 1972 I had a phone call and letter from a man who regularly has dried morels sent to him from Iowa, which he eats and enjoys throughout the year; he evidently cooks the dinners when he entertains. On this occasion he had prepared the morels as was his custom, and served them at dinner, with a highball beforehand. His guest became ill, with symptoms very much like those described above for *Coprinus*-alcohol poisoning. He then sent several specimens of the morels to me to see if there was possibly anything wrong with them — he thought that maybe they had been invaded by toxin-producing molds. They were not moldy (in fact I was able to get quite a number of cultures of living morel from them); the mycelium or spores, or both, in the caps were still alive. I also had them tested for bacteria and essentially none were found, certainly none that might cause food poisoning. These were good morels, no question about it. In looking through some of the old literature, I found a similar case described, in which a single person had developed symptoms of illness after consuming morels plus alcohol. These are the only two cases I know of involving toxicity of a morel-alcohol combination.

Hallucinogenic Mushrooms

The Genera *Psilocybe, Panaeolus,* and *Stropharia*

Since the early 1950s, hallucinogenic mushrooms and hallucinogenic compounds derived from them and from other fungi, principally ergot, have received a good deal of publicity, some of it of a pretty fantastic nature. One of these hallucinogenic drugs from ergot, and related to some of the compounds found in the *Psilocybe-Panaeolus-Stropharia* genera of mushrooms, has been referred to by an otherwise responsible author as the Drug of the Century. Supposedly it, or a derivative of it, will alleviate or totally eliminate many of the psychological maladjustments and traumas which keep us from achieving our full potential. Other equally enthusiastic and unsupported claims have been made for the mind-liberating qualities of some of the hallucinogenic mushrooms.

That some of these claims might be exaggerated is indicated by the experiences recounted by Schultes (16). Schultes is an ethnologist and botanist of high repute who, in 1940, published an article on hallucinogenic mushrooms of southern Mexico. Before this he had lived for some time with native Indians in Colombia and, in southern Mexico, he became interested in the use of hallucinogenic mushrooms. He identified the mushrooms as *Panaeolus campanulatus*, var. *sphinc-*

trinus, although, as will be seen shortly, several species in two or more genera may be similarly used. In the tribes that he visited there were professional divinators or witch doctors or medicine men who earned a living of sorts by locating stolen property, discovering secrets, and giving advice to their clients, even as other fortunetellers have done and still do in other societies, including our own, with or without benefit of trance-inducing drugs. Eating of *P. campanulatus* induced a semiconscious state in which the divinator saw visions in color, some of which were of kaleidoscopic patterns, and during which he gave out with what Schultes called "incoherent utterances." Consumption of fifteen fruit bodies of the mushroom (these are small mushrooms of little substance) produced the desired effects. Larger doses resulted in severe poisoning, and continued use of large quantities resulted in permanent insanity. Continued use of smaller quantities led to premature aging and early death.

Visions in color and incoherent and unintelligible jabbering can hardly be viewed by most rational people as evidence that these medicine men were in tune with the infinite; they obviously were out of tune with everyday reality, but that in itself is no justification for supposing that they were in tune with any celestial orchestration. They were deranged. Or, if you prefer, drunk. Certainly none of these native divinators, so far as I have read, have brought anything of value back from their drug-induced fits. The tribes still lived poorly and precariously.

Barron *et al.* (3) in an article in the *Scientific American* on mescaline, from the Peyote cactus, and psilocybin and psilocin, from *Psilocybe mexicana* and *Stropharia cubensis*, say that consumption of these can produce a wide range of subjective and objective effects, depending on the amount taken, on the personality and mood of the taker, and on the social milieu in which he lives. The visual effects are striking, especially when the eyes are closed; the mushroom eaters say they become one with the universe, whatever that means, if anything, and it is beyond their power to explain their visions. I still do not see how or why various changing patterns of lines and colors or a sort of glorious fireworks or the flashes of whatever might be seen through an internally mounted kaleidoscope are so extraordinary. What is there in it to "explain"?

Wasson (20) in 1959 summarized in the *New York Academy of Sciences Transactions*, a reasonably sober journal, what he calls six years of "research" on the hallucinogenic mushrooms of southern Mexico and Guatemala; he had learned of the existence of these mushrooms in 1952, a good dozen years after Schultes had published his article on them. He says he is trying to avoid the extremes of dry fact and unbridled imagination, and he certainly succeeded in avoiding one of these extremes by a wide margin. He and his wife made annual trips to Indian villages in the remote backwoods that could be reached only by donkey. Few of the natives understood any Spanish, and they had no written language of their own, so one wonders whether the Wassons picked up enough of the lingo to communicate with them. He refers to the hallucinogenic mushrooms used by these people as sacred and divine, although Schultes, the professional ethnobotanist, did not so refer to them at all. Wasson said that 14 species of *Psilocybe* were involved, but he gives no technical description of them and so we have no basis for judging the validity of the species. In many genera of fungi, as in many genera of other living things, closely related species grade into one another by such small steps that it is a matter of personal prejudice where to place the limits of a given species; years of collection and study might be required to establish reasonable limits of the species. Wasson was an investment banker before he became an expert on mushrooms. I am continually amazed by the ease and rapidity with which those who, with no previous acquaintance with mushrooms, become expert in almost any aspect of the study of them, including taxonomy.

The Indians, according to this account (still presumably not skirting too near the perilous shoals of dry fact), feared and adored these mushrooms, regarded them as the key to communication with the deity, consumed the mushrooms only in the silence and darkness of night behind closed doors, did not traffic in them openly (Schultes had obtained some of the mushrooms very readily, in trade for a few quinine pills; the tribal customs must have undergone very drastic changes in a few years), and talked of them only in whispers and with trusted friends. (They had talked freely and openly with Schultes, the pro.) Maybe it doesn't make all that much difference, but I think it is important to distinguish between information and entertainment. If a student is seeking knowledge he has a right to expect that his teacher will not bamboozle him with romanticism in the guise of "research" and that the preceptor will make every effort to distinguish fact from fancy.

Wasson says that the medicine men or women after eating these mushrooms could predict the future with astonishing accuracy, although no actual example of such prediction is given — and if some examples were cited I would want them to be verified by a number of independent and critical witnesses. Anyone can predict the future or call up monsters from the deep, but the trick is in predicting the future accurately or in having the monsters come at our call. Wasson says that those under the influence of the mushroom drugs could locate lost or stolen property, and communicate with friends and relatives at a distance. Again nothing is said about the details of said communications. What did they consist of? Who said what to whom? He does not say whether consumption of these mushrooms might enable us to predict stock market fluctuations or to fill out income tax forms. I doubt it. It would be good to have some of this backed up by a competent professional, or preferably by several competent professionals, just to have some sort of cross-check. In our ordinary research work I say that a given truth is not established until it has been found out by half a dozen different research workers at different times and different places; otherwise it too often turns out to be what I disrespectfully refer to as "thesis biology." The same should apply here.

The Species *Amanita Muscaria*

Nearly all books on mushrooms describe *Amanita muscaria* as poisonous, or as deadly poisonous, although a few mention that it has been used by a tribe, or by several tribes, in Siberia as an intoxicant. Fischer devotes five pages to poisonings by *A. muscaria*: symptoms, toxins, treatment, and so on. He includes a brief description of the fatal poisoning by *A. muscaria* of a Count de Vecchii (another account spells it Vecchj) in November 1897 in Washington, D.C. — and refers to it as a "classical" case, to my mind something like referring to a really horrible automobile accident as a "classical" accident. The count had bought a mess of what he thought were *A. caesaria* from a friend who had gathered them in the woods in nearby Virginia; presumably this friend also thought that they were *A. caesaria*, which is an edible species common in Italy, and which resembles *A. muscaria*. The count ate two dozen specimens of the mushroom for breakfast (another gourmet, evidently) which he finished at 8:30. By 9:00 A.M. he collapsed on his bed, lost his sight, suffered spasms of rigidity, developed convulsions so violent that he broke down the bed, then lost consciousness, and died the next day. Another friend (not the collector-seller of the mushrooms) who shared the mushrooms with him, but who ate only a dozen, developed double vision, became unconscious and remained so for several hours, but experienced no pain, and, after being given medication, recovered. According to Fischer the clinical symptoms of poisoning by *A. muscaria* include excessive salivation and perspiration, nausea, retching, vomiting, watery diarrhea, giddiness, confusion of ideas, hallucinations which resemble those of alcoholic intoxication, and delirium followed by loss or impairment of memory. This sounds rough and unpleasant.

Fischer also says that the use of *Amanita muscaria* to produce drunkenness is well known, and quotes a Russian, Krasheninnikoff, who traveled in Siberia and Kamchatka for ten years in the early 1700s (he must have been a hardy and adventurous soul) and reported that members of the Korak tribe used dried specimens of *A. muscaria* to produce intoxication — three or four specimens for a moderate sort of Saturday night binge, and ten specimens for a really good drunk. A single specimen was bought by them for what then was $20 worth of furs. Some of those who ate the mushrooms experienced a "horrible kind of delirium" and visions. Sometimes the intoxication was prolonged or passed on by drinking the renal secretion, and in this way a spree could be economically kept up for a week. Renal secretion, in plain English, is urine. Evidently those who kept up the party by drinking the renal secretions of their drunken companions were not considered to be among the socially elite of the tribe, at least by outsiders. The intoxication must have given them at least a temporary, if brutal, release from reality.

The same article by Wasson as was cited above, in which he was so anxious to avoid the extremes of unbridled imagination and dry fact, especially the extreme of dry fact, later says, "What was our surprise and delight to discover that there are people in Siberia who have worshipped a certain kind of mushroom from antiquity to our own times." None of the several people who up to that time had written about the use of *Amanita muscaria* to produce intoxication had even remotely suggested anything akin to "worship." I suppose it depends on what one means by "worship," but if the word is used in the ordinary dictionary sense — to pay divine honor to; to reverence with supreme respect and veneration — then what is described above can hardly rate as worship, any more than the rowdies who drink and fight and fall down on their faces in the alley behind the corner bar on Saturday night can be said to be worshiping.

Wasson and Wasson (21) even wrote a book, *Mushrooms, Russia, and History*, which retails for $125.00 (the investment banker influence probably shows up here), on these and other "sacred" mushrooms. How was it known that, since ancient times, mushrooms had been revered as sacred? Elementary: the Greeks had a proverb, "Mushrooms are the food of the gods." If this seems like rather obscure and tenuous "proof" of the sacred nature of mushrooms, it evidently satisfies Wasson and Wasson's requirements. Not only that; there is another book, by one Allegro (2), a philologist, entitled *The Sacred Mushroom and the Cross* in which it is established, chiefly by word comparisons, that Judaism, Christianity, and goodness knows what other religions, past and present, actually had their founding in *Amanita muscaria*. He also seems to revel, or maybe wallow would be a better word, in a mass of detail comparing the emerging cap of the young mushroom with the male penis, and the split volva (actually *A. muscaria* is one species of *Amanita* that doesn't have much of a volva) with the vulva of the female. I suppose it is beside the point that the book is dreadfully dull. I would rate the whole theme as remotely possible but highly improbable.

Other Hallucinogenic Mushrooms

From a 1973 book, *Hallucinogens and Shamanism*, chapter 7, "The Mushrooms of Language" (9), comes this comment:

> The Indian shamans are not contemplative, they are workers who actively express themselves by speaking, creators engaged in an endeavor of ontological, existential disclosure. For them, the shamanistic condition provoked by the mushrooms is intuitionary, not hallucinatory.

What one envisions has an ethical relation to reality, is indeed often the path to be followed. To see is to realize, to understand. But even more important than visions for the Mazatec shaman are words as real as the realities of the real they utter. It is as if the mushrooms revealed a primordial activity of signification, for once the shaman has eaten them, he begins to speak and continues to speak throughout the shamanistic session of ecstatic language. The phenomenon most distinctive of the mushrooms' effect is the inspired capacity to speak. Those who eat them are men of language, illuminated with the spirit, who call themselves the ones who speak, those who say. The shaman, chanting in a melodic singsong, saying *says* at the end of each phrase of saying, is in communication with the origins of creation, the sources of the voice, and the fountains of the word, related to reality from the heart of his existential ecstasy by the active mediation of language: the articulation of meaning and experience.

This is gibberish! A somewhat different and certainly more rational account of the effects of some of these hallucinogenic mushrooms is given by Sanford (15) in an article entitled "Japan's 'Laughing Mushrooms,'" a name given to the species *Panaeolus papilionaceus* because those who consumed them were soon overcome by an uncontrollable urge to laugh, and sometimes also to dance or jump around. The same mushroom occurs throughout much of the United States, and Sanford says that it was gathered and eaten by the thrifty farmers of Maine to give them a free drunk. He also quotes from the symptoms experienced by a botanist who shared a mess of these mushrooms with the wife of a friend at dinner; the lady's husband was at dinner too, but evidently did not eat any of the mushrooms. A few excerpts to end this chapter:

They were all eaten by Mrs. Y. and myself. Peculiar symptoms were perceived in a very short time. Noticed first that I could not collect my thoughts easily, when addressed, nor answer readily. Could not will to arise promptly. Walked a short distance; the time was short, but seemed long drawn out; could walk straight but seemed drowsy; remember little about the walk. Mrs. Y. was in about the same condition, according to Mr. Y. My mind very soon appeared to clear up somewhat and things began to seem funny and rather like intoxication. Walked with Mr. Y. A little later objects took on peculiar bright colors. A field of redtop grass seemed to lie in horizontal stripes of bright red and green, and a peculiar green haze spread itself over all the landscape. At this time

Mrs. Y. saw nearly everything green but the sky was blue; her white handkerchief appeared green to her; and the tips of her fingers seemed to be like the heads of snakes.

Next, say about half an hour after eating, both of us had an irresistible impulse to run and jump, which we did freely.

After entering the house, I noticed that the irregular figures on the wall-paper seemed to have creepy and crawling motions, contracting and expanding continually, though not changing their forms; finally they began to project from the wall and grew out toward me from it with uncanny motions.

I then had a very disagreeable illusion. Innumerable human faces, of all sorts and sizes, but all hideous, seemed to fill the room and to extend off in multitudes to interminable distances, while many were close to me on all sides. They were all grimacing rapidly and horribly and undergoing contortions, all the time growing more and more hideous. Some were upside down.

He pretty obviously was out of tune with reality for a while, but whether he was in tune with anything else seems questionable. The Great Secret was not revealed; the Vision remained murky.

2

Ergot and Ergotism

Ergot has plagued some of man's crops and ergotism has tortured man's body and the bodies of some of his domestic animals ever since he began to cultivate various species of grasses for their edible seeds. The human plagues from consumption of ergot probably are a thing of the past, but not the very distant past, since there were outbreaks of ergotism late in the nineteenth century, and even a few after 1900. Ergot itself, however, still is very much with us, as you will see; it is present regularly every year on wheat, rye, and wild rice (the ergot on wild rice is unusual in that it produces few or no toxic alkaloids), and there have been at least moderate epidemics of ergot on wheat and rye and on the wheat-rye hybrid triticale within the last few years in the upper midwestern states of the United States and in the Prairie Provinces of Canada. Proponents of organic foods object to the removal in the modern milling process of what they consider to be essential natural portions of wheat and their replacement with "artificial" or at least highly refined ingredients. However, one of the things that modern grain processing removes from wheat and rye before these are ground into flour is ergot, and it is thanks to this that ergotism in the human population is pretty much a thing of the past, and is likely to remain so.

Ergot has been more than just a scourge — we have got a lot of benefits from it, too, mainly in medicine; one investigator says that ergot has been found to be a veritable "treasurehouse" of physiologically (and psychologically) active chemicals. Over the years, many books have been written on various aspects of ergot — the last one, to my knowledge, in 1970 — and there have been thousands of research papers published on the chemistry and chemicals and other aspects of ergot; in the last 40 years these papers have appeared at an increasing rate, evidence that ergot still is very much in the scientific news. Doubtless it will continue to be, and this, as well as the many interesting aspects of the general biology of ergot, seems to me to justify the summary here presented.

The Fungus and Its Hosts

Ergot (the word is applied both to the fungus and also to the disease that the fungus causes) is a fungus, with the generic name of *Claviceps*, which means "club head," a name much more apt than many of those coined by taxonomists. The genus *Claviceps* has many species — anywhere from 30 to 50 or more, depending on which authority you follow; as with many other genera of fungi or insects or what have you, what constitutes a species often is a matter of personal persuasion or of personal revelation. In any case, there are quite a few species of *Claviceps*. They parasitize many species of the grass family — which includes all of our cultivated cereals such as wheat, rye, oats, barley, corn, and rice, and even sugarcane, plus many species of pasture and hay grasses, and many wild grasses, including bamboo, which is also a grass. Some species of *Claviceps* have a wide host range; *C. purpurea*, the common ergot on rye and wheat, for example, can infect more than a hundred species of wild and cultivated grasses — one author says "several hundred" but I am unable to find any confirmation of this — whereas other species of *Claviceps* can infect only one or a few host species. The one on wild rice appears to be restricted to wild rice.

Life History of the Fungus

The following account may be a bit dull, as life histories of fungi are likely to be, but it is not particularly devious or difficult, and familiarity with it will help furnish a basis for understanding some of the discussion to come.

Regardless of the species of *Claviceps* involved, or the host it attacks, the life history is pretty much the same. The fungus lives over winter in the form of a sclerotium (the term means ''hard''), a dense mass of fungus cells, on the surface of the ground. Figure 2-1 illustrates a sclerotium, or sclerotia. Within a given species of *Claviceps* the sclerotia will be somewhat similar to one another in size and shape, but among

Figure 2-1. Ergot sclerotia in rye (left) and wheat

the different species there is considerable diversity in shape, from spherical to elongate. Usually the sclerotia are somewhat larger and also less dense, or lighter, than the seeds of the host plant on which they are borne, which is fortunate, since this permits relatively easy separation of at least most of the sclerotia from the seeds by simple screening or flotation. The sclerotia have to undergo a period of dormancy, of anywhere from a few weeks to a few months, and sometimes also a period of low temperature, before they will germinate. That is, they are adapted to passing through the winter in a dormant state and do not germinate until their host plants, in whatever geographical region the host plants and the fungus occur, are flowering in the spring. In the north temperate zones, this means that the sclerotia are exposed to winter and rough weather for some months, the dormant period terminating during a time when there are irregular alternations of warm and cold and wet and dry weather; in more southerly climes such as Spain, Italy or the southern region of the United States, the dormancy may last only a few weeks, with no really low temperatures.

For class demonstration of germinating sclerotia, I have collected the sclerotia of ergot from Minnesota rye in the fall and have put these on the surface of moist sand, then have put them in an incubator at 4°–5° C (40°–42° F) and have left them until spring, at which time they were exposed to outdoor weather; after a few weeks, they began to germinate, as shown in figure 2-2. If the sclerotia are kept moist and at 3°–4° C (about 40° F) for a couple of months, then held at 14° C (57° F), they will germinate by mid-December. By manipulation of the temperature-time schedule, that is, the sclerotia can be induced to germinate at various times, but in nature they germinate when it is their time to germinate — when their host plants are flowering.

Each of the germinating sclerotia sends up several to many slender stalks, on the tip of each of which a spherical head (whence *Claviceps*) develops. Just beneath the surface of the head numerous cavities, called perithecia, are formed, and within each perithecium some thousands of asci grow out from the wall; each ascus bears eight long, narrow ascospores. These ascospores are shot out, to be spread by the wind, or ooze out, to be spread by insects, mainly small flies. Several successive crops of asci and ascospores may be produced in each perithecium, if the

Figure 2-2. Germinating ergot sclerotia. Ascospores are produced in perithecia just beneath the surface of the spherical heads.

weather is favorable, so that for a short time these ascospores may be relatively abundant. Some of these spores land in or are carried into flowers of the host plant. Here they germinate and produce mycelium, and this mycelium replaces the ovary of the flower. Within seven to eight days this mycelium in the infected flower produces a crop of millions of small, one-celled spores, the conidia, along with a sweet liquid, called honeydew. The honeydew may be produced in such quantity that it oozes out between the glumes, laden with conidia. This conidia-laden honeydew attracts insects, especially minute flies that visit grass flowers for their own small purposes, and, if the weather is humid and the temperature is right, the fungus is rapidly spread to other flowers of the same plant and to the flowers of nearby plants. Thus the first, or primary, infection by ascospores usually is limited to a relatively small percentage of flowers on a small percentage of susceptible host plants in a given area, but the subsequent infection, from insect-borne conidia, may involve a much larger number of flowers on a much larger number of plants. Given favorable conditions in the way of moisture and temperature, an epidemic can result.

Once the fungus is established in the flower, it may merely abort the flower and proceed no farther. Alternatively, if all goes well for the fungus, it grows apace and forms the spherical or elongated seed-shaped mass that, at maturity, hardens into the characteristic sclerotium, with a black or deep purple and slightly furrowed surface, and a white or pink-tinged interior. These sclerotia may be eaten by foraging animals, or may fall to the ground and lie dormant until the following spring, or may be harvested with the grain or seed. In the normal course of events, some of the sclerotia follow each of these three routes.

Those sclerotia which are harvested with the grain may be ground into flour and baked into bread — which in the old days gave rise to the horrid epidemics of ergotism to be described shortly — or they may be removed by screening or by other means, in which case they may end up in the screenings along with weed seeds, miscellaneous plant debris, insect fragments, small stones, and so on, and be ground into feed for livestock and poultry. The sclerotia may also be separated from the grain and sold for the drugs to be extracted from them. At times, in fact, the ergot harvested from a field of rye has had a much higher cash value than the rye from the same field. More about that later.

The essence or synopsis of the life history of *Claviceps* is that the fungus lives over winter in a dormant state, produces in the spring a crop of ascospores, the so-called sexual spores, that result in primary infection, and this is followed in about a week by the production of conidia that cause the secondary infection. The life history is not too much different from that of about 10,000 other species of *Ascomycetes* that have developed on the same branch of the evolutionary tree.

One qualification that must be attached to the account above is that it describes the general plan only, not the manifold deviations that may be encountered among the many species of *Claviceps*. As with most other kinds of living things, so with *Claviceps*, there are within each species an infinity of varieties, races, biotypes, and strains that differ from one another in morphology, chemistry, physiology, geographic distribution, and so on. In other words, *Claviceps* is biologically diverse, a characteristic shared by most kinds of living things. Darwin said, "Every species is varying in all characters at all times and everywhere." In Genesis 1:11 it is stated, "And God said, Let the earth bring forth grass'' (ergot is not mentioned) ''*and* the fruit tree yielding fruit after his kind . . .'' The ergot fungus certainly has brought forth a bewildering variety of ergot after its kind. All of them are ergot, to be sure, and to that extent they are ''of their own kind,'' but they are far from identical; *Claviceps* is dynamic, not static.

Chemical Makeup of Ergot

As mentioned earlier in this chapter, ergot sclerotia have been said to be a chemical ''treasurehouse,'' containing hundreds of complex chemical compounds. Bovè (23), with perhaps a touch of journalistic fervor — he is a member of the National Association of Science Writers, Inc. — says that this is ''Unbelievable. Really and truly unbelievable.'' It is not really all that unbelievable, especially since it happens to be true. A mustard seed contains hundreds of chemical products, and so do pine needles and probably just about every other living thing, even the most simple. Common molds and mushrooms contain a wealth of chemical compounds too, some of which will be taken up further on, and the investigation of a few of these is going on apace. When organic chemistry began to flower, more than a century ago, ergot already was known to contain compounds of medical and pharmacological interest, and so it was only natural that it should be investigated thoroughly, as it has been. Gröger (25) says, ''Probably only a few other drugs have been so thoroughly investigated chemically as has ergot.'' There is even a book devoted entirely to ergot alkaloids, *Die Mutterkornalkaloide*, by Albert Hofmann, published in 1964.

As indicated by the title of Hofmann's book, the compounds of greatest interest in ergot probably are those known as alkaloids. The dictionary defines an alkaloid as ''An organic substance having alkaline, or basic properties; now usually restricted to such as occur naturally in plants and animals. The alkaloids usually occur combined with some common organic acid, as tannic, malic, and citric. All contain nitrogen, carbon, and hydrogen, some being amino derivatives of aliphatic hydrocarbons.'' This is not very informative to the layman, unfortunately. If you really want to know what alkaloids are, you have to learn some chemistry. The dictionary continues, ''The alkaloids as a class have a bitter taste and in many cases are very poisonous. Some are important drugs, as morphine, quinine, cocaine, etc." Gröger says that alkaloids are the most important substances in toxicology and medicine, an opinion that I suppose one could find fault with, but at least it helps explain why ergot alkaloids are of interest.

Ergotamine, the compound used to hurry normal labor in childbirth, and to reduce hemorrhaging, was first isolated in pure and crystalline form in 1918. Lysergic acid, a breakdown product prepared in the laboratory from ergot, was first isolated in 1934, and a derivative of this, lysergic acid diethylamid, was made in a laboratory in Switzerland in 1943. This, of course, is the now famous, or infamous, LSD, the initials of which come from the German-Swiss name of the compound, Lyserg-Säure Diaethylamid, Säure being German for acid. Anyone especially interested in the detailed chemistry of ergot is referred to the book by Bovè and the summary article by Gröger, which is devoted chiefly to work since 1950, and in which nearly 300 references are cited — this gives some idea of the attention being paid to the chemistry of ergot at the present time, especially to the chemistry of ergot alkaloids and their derivatives.

The physiologically active alkaloids in ergot sclerotia, or at least in some sclerotia of some species of ergot at some times and places, are by no means unique. Similar compounds are produced by some common molds, by some common and uncommon mushrooms, such as those used for their hallucinogenic properties by ancient Indians of southern Mexico and Guatemala, as described in chapter 1, and also are found in the seeds of several species of higher plants in the family Convolvulaceae — the morning glory family — which has close to a thousand species. This wide but random distribution of similar and physiologically active alkaloids through the plant kingdom probably is not so wonderful as it seems; it is just another bit of evidence of the basic relationships of all living things. Many other chemicals and chemical processes are distributed in this fashion through the plant kingdom, and even through both the plant and the animal kingdoms. Some of the chemicals regulating some aspects of sexual activity and sexual differentiation, for example, are very similar in some of the fungi, in rodents, in pigs, and in man, and a chemical now used to alleviate postmenopause miseries in women is derived from an estrogenic compound that is produced by a common fungus, *Fusarium roseum*, which will be met again in chapter 4. Why shouldn't some of the processes of life be similar in fungi, pigs, and man — presumably all were derived by slow evolution from the same primordial ooze, and some of the vital processes got carried along into various branches and twigs of the evolutionary tree and became, so to speak, isolated here and there as a casual or vital part of the economy of this or that organism.

The compounds of interest and importance in ergot occur in different amounts in sclerotia gathered from a single host species growing in different places in the same season, in sclerotia from different species of plants growing in the same area in the same season, and in sclerotia from the same species of plant growing in the same place in different years. This is true of other compounds produced by other plants — it is, in fact, the rule rather than the exception. A given variety of grapes, for example, growing on a given slope in a given vineyard in a given year, may produce a unique wine, with its own body, bouquet, and other esoteric real or imaginary characteristics on which a wine is rated. Any wine connoisseur knows that as a matter of course, as do also those who sell the wine, down to and including the clerk in the local liquor store who does not even know how to pronounce the name of either the wine or the region where it was grown. The same differences occur, but probably to a lesser extent, in soybeans, wheat, and potatoes, but few gourmets become rapturous over the provenance or the vintage of a helping of soybeans, and the waiter does not present the diner with a potato or a loaf of bread cradled gently on a napkin, as he does with a bottle of wine. It just is not customary to perform this stylized ritual with anything but wine. In any case, these differences are just another example of the infinite diversity among the species of living things — they are alive, constantly varying, never exactly the same, always changing, and are not identical replicates punched out on a press.

Ergotism

In Man

Some writers have claimed that ergot and ergotism were common, or at least that they were encountered occasionally, in ancient Greece and Rome, but this is based on tenuous evidence in old writings in which the words that may have referred to ergot or ergotism may also have referred to something quite different. Rye was not grown in Greece, and it probably did not occur there as a common weed grass in fields of other cereal crops. Ergot, of course, occurs on many grasses other than rye, but again there is no evidence that in Greece seeds of these other grasses were consumed in any quantity as food.

Rye supposedly was not grown deliberately for food until some time in the early Middle Ages, in what is now eastern Europe and western Russia. There, rye occurred naturally as a weed in fields of wheat, and as cultivation of wheat spread across northern Europe, the wheat sometimes was supplanted by rye, intentionally or unintentionally, in part because rye was better adapted to the cool, moist climate that prevailed. With the increase in people came an increase in rye, and also an increase in ergot and in ergotism.

Even in those days at least some of the people recognized that ergot, as it developed on the heads of rye, was of more or less foreign nature, and there were many mostly fanciful explanations to account for its presence. That it was common, at least at some times and in some places, and that the people recognized it as something other than rye, is indicated by the many names applied to it as it occurred on plants in the field — Bovè lists 45 names in French, 62 in German, 21 in Dutch, 9 in English, and 48 more in 16 other languages.

A portion of the ergot on rye was harvested with the rye kernels and was ground up into flour, made into bread, and consumed. Even with modern harvesting and cleaning equipment, some ergot occasionally ends up in our rye or wheat flour, or in semolina. Consumption of even relatively small amounts of ergot over a period of weeks, if the ergot contains an appreciable amount of ergotamine, can result in a variety of miseries, and consumption of larger amounts over a relatively short time — one or more meals, depending on the amount of toxin present — can also cause severe illness. According to Barger (22), in some of the epidemics of ergotism in France and Germany hundreds of years ago, the bread contained by actual measurement from 2 percent to 15 percent of ergot, an almost fantastically large amount. The Canadian Department of Agriculture has established a maximum limit, or tolerance, of 0.1 percent of ergot in animal feed that may be sold in

Canada. So in some of the European outbreaks of ergotism the people were consuming from 20 to 150 times as much ergot as is permitted in animal feed in Canada. As will be seen below, this limit of 0.1 percent probably is too high for safety, even for cattle.

From the year 900, when the keeping of records evidently became common in many regions in France and Germany, to around 1300, there were severe epidemics of ergotism over large areas every five to ten years. Most likely there were lesser ones here and there each year. Between 922 and 1254 A.D., 23 such outbreaks were recorded. In the epidemic of 944 in southern France, it is said that 40,000 people died. The cause was unknown, no cure was available — you do not *have* to know the cause of a plant or animal disease in order to cure it, but often such knowledge helps and, with ergot, the only sure preventive was to avoid consumption of ergot. Until people knew that consumption of ergot was the cause of their sufferings, they had no rational method of treatment. So, as a possible cure or palliative, the bones of various saints were displayed to the stricken multitude — different saints in different places. Eventually St. Anthony became so associated with ergotism that the affliction became widely known as St. Anthony's Fire. The fire referred to was the burning sensation experienced by many of the sufferers. The situation must often have been horrible. A few quotations from Barger, who in turn quotes predecessors, who in turn may be quoting their predecessors, will illustrate some aspects of the outbreaks of ergotism:

In 944 there was a violent epidemic in Aquitaine and Limousin. It is mentioned in half a dozen chronicles one of which states that over 40,000 persons died by this pestilence. (Perhaps the writer confused it with bubonic plague.) After an extremely severe winter there followed a great drought and scarcity. The end of the millennium was approaching and when a plague of invisible fire broke out, cutting off limbs from the body and consuming many in a single night, the sufferers thronged to the churches and invoked the help of the Saints. The cries of those in pain and the shedding of burned up limbs alike excited pity; the stench of rotten flesh was unbearable; many were however cooled by the sprinkling of holy water and snatched from mortal peril. It is evident that the epidemic occurred during several months following the harvest. . . .

In 1039, or perhaps a few years later, a "deadly burning" consumed many of all classes. It was regarded as a sign of the divine wrath for the breaking of the truce of God (which, *inter alia*, restricted fighting to Mondays, Tuesdays, and Wednesdays). The sins of the princes were visited on their peoples; some patients survived in a mutilated condition as an example to those coming after. [Thus the poor peasant was not only put upon and woefully mistreated by the prince and his henchmen, but also was punished for the prince's sins. What injustice!] . . .

Both the convulsive and the gangrenous symptoms are explicitly mentioned in the next account [Lorraine, 1085]. "Many were tortured and twisted by a contraction of the nerves; others died miserably, their limbs eaten up by the holy fire and blackened like charcoal." Here for the first time the fire is called "holy." . . .

Among the many accounts, the most detailed description of the disease is contained in a book on the miracles of St Mary of Soissons, *de curatione ardentium*, by a contemporary, Hugh Farsit. He calls the disease a wasting one . . . Under the stretched and livid skin the flesh was separated from the bones and consumed. Death was denied to the sufferers until the fire invaded the vital organs. Most strangely this fire could consume without heat, and poured over the sufferers such an icy cold, that they could not be warmed by any means; and what was no less strange, if by divine grace, the fire had been extinguished, so great a heat pervaded the limbs of the sick, that it was often accompanied by cancer (gangrene?) unless treated by medicaments.

Poisoning by ergot was not alleviated by "medicaments." It still isn't. Barger continues: "The Marburg faculty gave the first detailed description of convulsive ergotism. The cries of the sufferers could be heard in villages 'beyond the eighth or the tenth house and quite far off in the fields.' " This was in 1597, and "the Marburg physicians were in error in considering the disease to be infectious, a belief shared by some later writers, no doubt because often in a family several members were attacked who would naturally live on the same diet. The exact cause of the disease remained as yet unknown and the Marburg faculty merely attributed it to bad food in general." That is, after widespread and repeated epidemics of ergotism over a period of nearly 1000 years, the true cause still eluded the physicians of the day — although some patients recovered if taken to a "hospital" in time, where they were given food free of ergot.

In 1717 there was an epidemic about which Barger quotes an authority as follows: "Severe cases differed only from true epilepsy in that patients were conscious. *Often three-quarters of the grain consisted of ergot and other impurities* [my italics]. It was even the subject of a printed sermon (Bruno) and of a theological thesis (Kunad); a belief in witchcraft was still prevalent and many believed the sufferers from convulsive ergotism to be possessed by demons."

Later Barger comments, "The inability of many physicians of this time [about 1770] to see in ergot the cause of a disease is no less remarkable than the persistence of the peasants in eating bread made from ergotised rye." The peasants' "persistence in eating bread made from ergotised rye" may not have been so remarkable; rye never, up to then and for a long time afterwards, was pure in the sense of consisting of only sound rye kernels; it contained various amounts of various and sundry extraneous materials in the way of dirt, stones, plant debris, chaff, and weed seeds, including those of darnel, or *Lolium*, which are toxic in their own right. It always had been that way. It still is. The nobility and gentry and some of the clergy often had their grain cleaned to some extent, but the peasants ate what they had available to eat, and the presence of a modicum of dark and bitter ergot in bread that, by nature, was dark and bitter anyway was no cause for alarm. Even today organic food enthusiasts consume, unknowingly but enthusiastically, some grain products containing rodent excreta and hairs, insects and insect fragments, and at least small quantities of ergot. Presumably they relish it.

In general, there seems to have been two types of ergotism, convulsive and gangrenous. In the gangrenous type, again quoting from Barger:

> The patient often began by complaining of a general lassitude, vague lumbar pains, or pains in a limb, particularly in the calf. The pulse and appetite remained at first normal; sometimes there was slight vomiting. The intellect was dulled. In the course of a few weeks the part affected (more often a foot than a hand) became somewhat swollen and inflamed, and was attacked by violent burning pains, as if "un fer ardent traversait le membre affecté." Hence the [name] of fire . . . A feeling of intense heat alternated with one of icy cold. Not being able to bear the heat in their beds, the sufferers would seek relief in the open air, and then feel so cold that they immersed their limbs in hot water Gradually the part affected became numbed; the pains sometimes stopped

suddenly. . . . Later the diseased part became black ("like charcoal," as the chronicles have it), often quite suddenly, and all sensation was lost. The gangrenous part shrank, became mummified and dry; the whole body was emaciated and the gangrene gradually spread upwards; sometimes there was putrefaction (moist gangrene).

In severe cases the course of the disease was much more rapid; with violent pains for twenty-four hours, as the only premonitory sign, gangrene might set in suddenly. The separation of the gangrenous part often took place spontaneously at a joint without pain or loss of blood. It is related that a woman was riding to the hospital on an ass, and was pushed against a shrub; her leg became detached at the knee, without any bleeding, and she carried it to the hospital in her arms.

On convulsive ergotism, again from Barger, but attributed by him to a translation from an earlier account in Latin:

It seized upon man with a twitching and kind of benummedness in the hands and feet, sometimes on one side, sometimes on the other, and sometimes on both: Hence a Convulsion invaded men on a sudden when they were about their daylie employments, and first the fingers and toes were troubled, which Convulsion afterwards came to the arms, knees, shoulders, hips, and indeed the whole body, until the sick would lie down and roul up their bodies round like a Ball, or else stretch out themselves straight at length: Terrible pains accompanied this evil, and great clamours and scrietchings did the sick make; some vomited when it first took them. This disease sometimes continued some days or weeks in the limbs, before it seized on the head, although fitting medicines were administered; which if they were neglected, the head was then presently troubled, and some had Epilepsies, after which fits some lay as it were dead six or eight hours, others were troubled with drowsiness, others with giddiness, which continued till the fourth day, and beyond with some, which either blindness or deafness ensued, or the Palsie: When the fit left them, men were exceeding hungry contrary to nature [another author tells about three men in Germany, suffering from ergotism, who, when they recovered from this first "fit" were so ravenous that each consumed *three pounds of bread* without stopping!]; afterwards for the most part a looseness followed, and in the most, the hands and feet swell'd or broke out with swellings full of waterish humours, but sweat never ensued. The disease was infectious, and the infection would continue in the body being taken once, six, seven, or twelve moneths. [That is, if the disease once was contracted or "taken" it would persist for six, seven, or twelve months. Or longer.]

> This disease had its original [he means origin] from pestilential thin humours first invading the brain and all the nerves; but those malignant humours proceeded from bad diet when there was scarcity of provision.

So, for close to a thousand years, ergotism was among the great pestilences of Europe and of what is now the USSR. There is no way to know, of course, just how much of the miseries ascribed mainly or solely to ergot actually were caused by consumption of ergot, and how much was the result of a combination of ergot poisoning with other afflictions. Many other diseases must have been constantly endemic over large areas, and occasionally epidemic, such as plague, typhus, tuberculosis, influenza, amoebic and bacillary dysentery, smallpox and other poxes, small and great, plus heavy intestinal infestations with roundworms, tapeworms, and hookworms. One case, indeed, is mentioned several hundred years ago in which the attending physician observed masses of roundworms in the stools of his patients suffering from ergot poisoning. This was in Germany, but the same situation must have prevailed in many places. The poor peasants often were huddled together, ill fed, dirty, ridden with lice, fleas, bedbugs, and other assorted vermin. As an example of what lice alone can do to an otherwise healthy farm animal, even when the lice are not carrying disease-causing agents, as they often do, a case of extremely heavy infestation of lice on a farm animal was described in the midwestern United States in 1973. Mind you, this was not medieval times, but 1973, and on a modern farm. One cow in a herd began to act in a wild manner, jumping and prancing and butting and bawling. The veterinarian was called, but even with help he could not corral the beast for a close examination. The cow soon died, and upon being examined was found to be densely covered with blood-filled lice, which had literally bled the poor animal to death; it was autopsied immediately, and even the heart was empty of blood: the lice had drained the cow of blood completely. Of course it is unlikely that any person even in medieval times could have been as heavily infested by lice as that, but they all were infested to some extent, as well as being invaded and infested and infected by a host of other parasites and predators; poisoning by ergot was an added misery to which they were exposed.

The epidemics decreased after 1600, but did not altogether disappear.

Bovè mentions 65 epidemics of convulsive ergotism between 1581 and 1889 — 29 in Germany, 11 in Russia, 10 in Sweden, and others in various other countries, including the United States. There must have been many lesser outbreaks of ergotism that went unreported: neither the diagnoses nor the public health records were all that good. There was a fairly extensive outbreak of ergotism in the USSR in 1926, and a lesser one in England in 1928–1929, the latter one among Jewish immigrants from eastern Europe who ate bread made from rye that was not properly cleaned to remove the ergot sclerotia. Just as a matter of interest, it might be noted that this rye, and of course the ergot in it, were stone-ground. Some food faddist stoutly maintain that stone-ground flour is more wholesome than flour ground with steel rolls, the stones preferably being driven by water power, but even stone-ground ergot is not wholesome.

In 1951 there was an outbreak of "bread poisoning" in a small town in southern France where, in medieval times, ergotism repeatedly had scourged the people. In the 1951 outbreak not only people, but cats, dogs, poultry, and fish were affected. The newspapers and radio broadcasts immediately made a sensation out of this — St. Anthony's Fire had returned! There was even a book written about it, somewhat belatedly — *The Day of St. Anthony's Fire* — by a New York "science" writer, who, 15 years after the event, visited the place, interviewed most of those concerned, including some of the dogs, cats, and fish, and wrote a really lurid firsthand account that established, at least to his satisfaction, that ergot poisoning by a peculiarly virulent form of the fungus not seen before or since was responsible. He did not, so far as is known, bring back any of this superpotent ergot. Bovè says that the toxicologists who studied the matter, and whose qualifications can be presumed to be somewhat better than those of a writer of fantastic science fiction, attributed the poisoning to a phosphorus insecticide applied to grain that later, by error, was ground into flour. This seems reasonable.

What are the chances of ingesting sufficient ergot in bread or other cereal products today to affect your health? Probably rather remote, depending in part on your life style. A few years ago I received samples of semolina — the coarse durum flour from which macaroni and spaghetti and other pasta products are made — that had been milled in northwestern Minnesota from durum wheat grown there. It was sent to me for examination because it contained small black particles that the customers objected to. Microscopic examination revealed that most of these black specks were particles of ergot sclerotia — not all of them, but most of them. Only the outer rind of ergot sclerotia is black; the interior is off-white, and so probably only 10 percent or so of the ergot fragments in the semolina were detectable as black specks. I do not know what percentage by weight of the semolina was made up of ergot, but I suspect that it was more than the 0.03 percent that is permitted in flour in some European countries where limits of ergot content in flour have been established. This ergoty semolina had been manufactured, sold, and consumed as pasta products for some months before complaints began to come in, so some people were consuming at least some ergot. Whether they consumed enough to affect them in any way seems doubtful, but then we do not know how much ergot has to be consumed by people to affect them.

In 1971 there was a fairly heavy infection of ergot in the wheat variety Waldron, then widely grown in the Red River Valley of Minnesota–North Dakota for milling into flour for bread. Before it was released this variety was known to be susceptible to infection by the ergot fungus, and the plant pathologists on the committee that decides on the release of new varieties for planting voted against its being released but were overruled. Waldron no longer is grown extensively, not because of its susceptibility to ergot infection, but because it has been replaced by newer, higher yielding, semidwarf varieties, some of whose genes come from the semidwarf wheats developed by Nobel Peace Prize winner Norman Borlaug, of the Rockefeller Foundation and the International Center for Wheat and Corn Improvement in Mexico. In 1972 there was enough ergot in some fields of durum wheat in Minnesota and North Dakota to be of concern to the millers, and there is some ergot in many fields of rye and of durum and spring wheats every year. So ergot is still with us.

We do not get enough ergot in commercially milled flour these days to be of any real concern, but the reason is that most of the ergot present in the grain at harvest is removed when the grain is cleaned at the country elevator, and most of the rest is removed when the grain is again cleaned at the mill. Concerning ergot, the *Official Grain Standards of the United States* (27) specifies that wheat or rye that contains more than 0.3 percent by weight of ergot shall be classed as "ergoty," and such "ergoty" wheat will receive a lower price because of the cost involved in removing the ergot. It is difficult to remove "all" of anything from grain, but in the case of ergot it must in practice be close to that; in the examination of many thousands of samples of wheat in grain storage work over the past 25 years or so, I cannot recall having seen so much as a single ergot sclerotium. This is where your life style, mentioned above, comes in. If you prefer or demand stone-ground or hand-ground flour made from organically grown wheat got directly from a farm, you or your miller probably should be able to recognize ergot sclerotia in the wheat you buy, and have some method or machinery to remove them.

In Domestic Animals

When epidemics of ergot poisoning scourged the human population, as described above, some ergot poisoning probably occurred in domestic animals too, although there are no records of this. In those times and places, when there was a shortage of good grain for human consumption, it us unlikely that much grain would have been fed to animals. However, if some of the ergot was separated from the grain itself, as apparently it was, it is probable that this portion, considered unfit for consumption even by peasants, was fed to the livestock, a practice that still is generally followed on the most up-to-date farms. As screens and screening procedures were developed to remove ergot sclerotia and other extraneous materials more efficiently from the grain, the problem of ergot poisoning in domestic animals probably became more acute — the improved cleaning methods removed more ergot from the grain and added it to the screenings, and the screenings were fed to livestock.

When Waldron wheat was moderately to heavily infected with ergot in the Red River Valley in 1971, our laboratory at the University of Minnesota received from a turkey grower a sample of screenings that contained more than 10 percent ergot; he wanted to know whether this amount of ergot in the screenings would, when added to the ration in the usual proportion, be harmful to the turkeys. We couldn't tell him, because no feeding tests have been made anywhere to provide the data necessary to answer this question. On the face of it this lack of data on what may be a very important problem in our agricultural economy may seem strange, especially since in our work on mycotoxins for more than a decade the major emphasis has been on feed-related problems in domestic animals. Why not get a few turkey poults, give them rations containing different amounts of ergot, and see what happens?

It isn't quite that simple. The amount and potency of the physiologically active toxins in ergot sclerotia vary from crop to crop and place to place and year to year, so the important thing is not the weight of ergot sclerotia consumed, but the amount of physiologically active alkaloids consumed. This can be determined, but it requires the service of a fungus physiologist or biochemist with the know-how and equipment and inclination and time to do it, plus fairly liberal financial support. To be meaningful, such tests should be made with rations containing graded amounts of ergot toxins fed throughout the life of the turkeys. Depending on the magnitude of the differences that are looked for, this might mean 40 to 50 turkeys on each of four or five different rations in addition, of course, to an equal number in the control group given feed known to be free not only of ergot toxins but also of other mycotoxins and of other harmful substances. Each group should be replicated at least once, and the whole test should be replicated in time — say each year for two or three years. The tests should be overseen by a veterinary pathologist and by a poultry nutritionist. By now we already have three rather high-priced professional investigators involved, and while this particular investigation would not require 100 percent of their research time, enough of it would be required so that they could not devote too much time to other things. The investigation also would require assistants and expensive laboratory equipment and supplies. Over a three-year period a budget of, say, $75,000 might do it. At the rate of inflation as this is being written in 1974, within a few years that amount could be doubled. Who is going to put up that sort of money? Not the turkey growers who want the answers.

In 1971 Dinusson *et al.* (24) reported the results of feeding tests in which beef cattle were given rations containing ergot from the wheat-rye hybrid called triticale. The *triti* part of this name comes from *Triticum*, the scientific name of wheat, and the *cale* from *Secale*, the scientific name of rye. This new and man-made genus has considerable promise as a feed grain, in part because of its heavy yields and its adaptation to areas where other feed grains such as maize and sorghum do not thrive. One of its drawbacks is its susceptibility to ergot infection, which is why Dinusson and his collaborators at North Dakota State University, Fargo, wanted to test the effects on animals of small amounts of ergot in the ration. They used ergot from wheat and rye, not from triticale, presumably because ergot from wheat and rye was more readily available than that from triticale. Also, they did not determine the amounts of specific ergot toxins present in the ergot they used, and so while we know the amounts of ergot in the ration we do not know the amounts of toxin or toxins in the ration. Some of their results are shown in tables 2-1 and 2-2.

In additional tests the results of which were not tabulated, 0.15 and even 0.06 percent of ergot in the ration resulted in lower than normal weight gain and in various other deleterious effects, notably an outbreak of footrot due probably to decreased flow of blood to the feet, digestive upsets, increased urination and wet pens, and lack of shedding of winter hair. They suggest that any rations containing 0.06 percent or more of ergot should be considered potentially toxic to beef cattle, especially for long-term feeding. This amount — 0.06 percent — is one-fifth of the amount of ergot permitted in grains before they are judged "ergoty" by the United States grain-grading standards.

In Canada, according to Seaman (26), no more than 0.1 percent by weight of ergot is permitted in feeds for poultry and other farm animals. Judged by the results summarized above, even this 0.1 percent may be

Table 2-1. The Effect of Ergot on Feed Intake and Gain of Beef Heifers

Gain and Intake (in Pounds)	Control	0.5% Rye Ergot	0.5% Wheat Ergot	0.1% Wheat Ergot
Initial weight	591	594	603	586
Final weight	623	592	572	571
Average daily gain	0.65	−0.04	−0.057	−0.31
Feed per day	11.1	8.88	8.00	7.77
Feed per pound gain	17.2			

Source: Adapted from Dinusson *et al.* (24).

Table 2-2. The Effect of Ergot on Feed Intake and Gain of Fattening Cattle

Gain and Intake (in Pounds)	Crossbreds		Beef Breed	
	Control	0.5% Barley Ergot	Control	0.5% Barley Ergot
Initial weight	538	520	382	372
Final weight	1,056	1,011	817	761
Average daily gain	2.19	2.12	1.88	1.68
Average feed per day	17.2	15.2	13.4	11.7
Feed per pound gain	7.8	7.2	7.1	7.0
Percentage of intake of control lot		88%		87%
Percentage of gain of control lot		97%		89%

Source: Adapted from Dinusson *et al.* (24).

too high for safety. And, although it is not so stated, this legal limit of 0.1 percent means "detectable ergot content." Detectable by whom, and when and how? Detectable in ground screenings after these are mixed with a dozen other ingredients into a given batch of feed? Would the man in charge of the feed mill at a country elevator sometimes deliberately mix ergoty grain into a batch of ground feed? Very likely he would, since the standards of behavior, or of ethics or what have you, of those in charge of making ground feeds probably is no lower and no higher than the standards of those not in charge of making ground feeds. Thin-layer chromatography can detect very small amounts of ergot present in some things, but not necessarily in all samples of ground feeds that contain such a variety of materials. Even if extraction, purification, and subsequent identification on thin-layer plates, followed by spectrographic analysis, does serve to detect very small amounts of the ergot toxins, this procedure is not one that can be used by the feed mill or the farmer. In any case, it is one thing to have a regulation on the statute books, and another thing to enforce it, and this is especially true concerning minimum amounts of undesirable but hard-to-detect ingredients in any mixed product. Also, partly as a matter of custom, partly from economic necessity, the purity or lack of it of ingredients in animal feeds has not been given much attention, and to some extent it can't be given much attention; one of the common ingredients of processed feeds, for example, is slaughterhouse offal, and even though it may smell to heaven and consist of what are to us aesthetically unattractive portions of the animal remains, it is basically a perfectly wholesome feed ingredient. As another example, just about all samples of hay that we have tested in our laboratory are moldy to some extent, and many of them are heavily moldy, and yet they seem to be accepted very well by cattle and horses, and usually cause no trouble. The economics of the dairy, beef, and poultry industries is such that anyone who insisted on feeds made only of ingredients of the highest purity probably would go broke in a month or less. So if the feed grain contains a little ergot, neither the farmer nor the feed processor is likely to get wrought up about it; feeding it is a calculated risk and, in farming, risk is the name of the game.

The problem of ergot toxicity in farm animals is not only one of feeding ergoty wheat or rye; *Claviceps purpurea* and other species of *Claviceps* infect many kinds of grasses, including some common pasture and forage grasses, and consumption of these may also result in poisoning. *Paspalum dilatatum*, commonly known as water grass, Dallis grass, paspalum, or paspalum grass, is an important forage grass in the southern United States and in various other countries around the world. One of its drawbacks, and occasionally a serious one, is that it is susceptible to infection by ergot, and evidently this ergot sometimes contains rather potent toxins. Consumption of the ergotized heads of paspalum by cattle results in the affected cattle losing the use of their forelegs; they fall down, and then starve or they fall in the water and drown. Essentially the same symptoms from cattle eating ergot-infested paspalum have been reported in New Zealand and Australia.

Foraging wild animals and wild seed-eating birds, including ducks and geese, must at times consume enough ergot sclerotia from wild and cultivated grasses to affect their health, but, if so, we know nothing about it and neither do those in state or federal wild life departments, since no one has ever so much as looked into the possibility. It would be interesting to find out.

Ergot and Its Derivatives in Medicine

A German book published in 1582 mentions ergot as a useful aid to promote labor in childbirth, and in the eighteenth century it was commonly used by midwives for this purpose. It was administered as a powder of ground-up sclerotia, or as a water decoction, but there was no standardized product, no way to know how much of the active ingredient or ingredients might be present in a given batch, and so it must have been pretty much of a guess how much to dose a patient with. Well, medicines are still prescribed to be taken "three times a day" because in ancient times the number three had magic powers or magic significance.

By 1880 ergot came to be accepted as a medically legitimate drug to promote labor in childbirth and to reduce postbirth hemorrhaging, and the dosage was standardized, after a fashion. That is, a given amount of an extract of ergot sclerotia containing the desired alkaloid, and probably containing also a host of other substances of unknown composition and unknown effects, was injected into a white leghorn rooster of given weight, and the time required for the rooster's comb to turn blue determined the amount of the active muscle-contracting principle present. From this, the dosage was calculated. Eventually the active principle for this use, ergotamine, was extracted in relatively pure form; a derivative of this, ergometrine, was developed by chemists and is now widely used.

In the 1920s ergotamine was found to be effective in the treatment of migraine headaches. Bovè says it still is the only specific drug for the alleviation of migraine headaches, and methysergile, another derivative of ergotamine, if given in time, prevents the onset of such headaches. Still another derivative gives promise, if tests with mice and rats prove out in humans, of being a very effective birth control drug — one pill per month taken in the second half of the menstrual cycle prevents the fertilized egg from becoming attached to the uterus wall. Work is under way on many other derivatives and on their physiological effects. Who knows what might be turned up.

Artificial Production of Ergot

With ergot in demand for pharmaceutical uses there has at times been a very good market for ergot sclerotia — if they contained enough ergotamine — and sometimes a farmer harvested his rye field for the ergot rather than for the rye. For a long time ergot sclerotia coming from Spain were considered to be superior to those from any other country. The astute merchants in Spain exploited this by importing ergot from other European countries into Spain, thus naturalizing it and increasing its value when shipped out. Some of them had another dodge — they molded imitation sclerotia out of dough and adulterated the genuine or pseudogenuine article with these, copying, in this, the clever Chinese who made imitation soybeans out of clay and sold them to the British to feed the donkeys working in the mines (the adulteration being discovered only after some of the donkeys fed in part on this clay soybean diet were found to be starving).

Numerous systems, devices, and means were tried to induce artificial epidemics of ergot in rye in the field. Chiefly these involved inoculation of the rye flowers with spores of *Claviceps*, either by direct injection by means of needles or by power sprays. Some of these procedures resulted in fairly heavy infection of ergot, followed by good crops of sclerotia — but the sclerotia had little or none of the desired alkaloids, and so were worthless. So far as I am aware, no one, in any of the several countries where such artificial epidemics have been worked on, has yet succeeded in consistently producing good crops of high-potency ergot, and the major dependence of the pharmaceutical industry for ergot still is on the naturally produced product. A firm in Dassel, Minnesota, has long been engaged in collecting ergot sclerotia from rye grown in the region, and selling it to the drug companies. Presumably it is a profitable enterprise but they have not become inordinately wealthy.

The ergot fungus — or the ergot fungi, since there are many species — can be grown in artificial culture, on agar or in a liquid medium in flasks or other vessels. A good deal of effort has been devoted to this, too, because if physiologically active alkaloids could be produced in this way it would be highly profitable — at least until someone else, a little smarter or luckier, came up with a better method. After all, the world market for ergot alkaloids is by no means very great.

Occasionally some isolates or strains of some species of *Claviceps*, grown in one or another liquid medium and under one or another set of conditions, have produced enough of one or another alkaloid to keep up the hopes of the researchers, and even the hopes of some of those subsidizing such research, who are likely to be much less optimistic than the researchers themselves. So far as I am aware, none of these attempts to produce medically valuable ergot alkaloids in culture have paid off financially; no one has yet reached that corner where success is just around.

3

Mycotoxins and Mycotoxicoses: Aflatoxin

Mycotoxins are toxic compounds produced by fungi. Technically, the toxins in poisonous mushrooms and in ergot are mycotoxins, too, but people poisoned by mushrooms or by ergot have to consume at least a moderate amount of the fungus tissues that contain the toxins, whereas in the mycotoxins discussed in this chapter and the next the major portion of the toxins is produced by the fungus in the substrate, or material, in which the fungus is growing. Some of the toxic compound or compounds may be present in the mycelium or in the spores of the fungus itself, but the amount of actual fungus substance consumed in cases of poisoning of this sort is likely to be infinitesimally small, and the material in which the fungus has grown and in which it has produced its toxin may not look or smell moldy. Antibiotics such as penicillin and streptomycin, produced by fungi, are mycotoxins too, but in general usage the term "antibiotics" is applied to compounds produced by microbes that are toxic to other microbes, whereas the term "mycotoxin" is applied to metabolic products produced by fungi that are toxic to higher animals, including man. Mycotoxicoses are the diseases resulting from ingestion of mycotoxins.

That some moldy foods and feeds might be injurious to the people or to the domestic animals that consume them is not at all a new idea. Before 1900 workers in Italy postulated a cause-and-effect relationship between consumption of moldy corn, especially by children, and the development of illness, including pellagra. Some of them even isolated strains of these suspect fungi from corn, grew them in pure culture, and from this culture material obtained compounds exceedingly toxic to animals. They did not, however, identify the compounds in question, nor did they detect them in the suspect corn. Japanese investigators shortly before and after 1900 established an association between consumption of moldy rice and the development of various diseases, including acute cardiac beriberi; the probability of this being a cause-and-effect relationship was claimed by Uraguchi (40) as late as 1969 to be reasonably high. In the United States and in some other countries there have been over the past hundred years or so occasional outbreaks of "moldy corn disease" in domestic animals, primarily horses, cows, and swine, and many cases of the "estrogenic syndrome" in swine were attributed to consumption of corn or other grain invaded by fungi, but in spite of abundant observational evidence to implicate fungi as a probable cause, there were no clear-cut results from controlled experiments that would eliminate other possible causes and pin the blame on one fungus. In the early 1950s there was an extensive outbreak of moldy corn disease in the southeastern United States; hundreds of swine foraging off harvested corn fields in the fall became ill, and many of them died. Investigation of the problem was undertaken by a couple of competent veterinary pathologists and a competent mycologist, who fortunately was interested primarily not just in the names of fungi but in what they did. A number of fungi were isolated from the suspect corn and grown on autoclaved moist corn, and this was combined into a ration and fed to pigs; one of the isolates of *Aspergillus flavus* when so grown and fed to pigs was rapidly lethal to them, and these pigs had outward signs and inward lesions similar to those in the field cases of moldy corn disease. This was good work, by able pros, and the results were published by Burnside *et al.* (29) in a veterinary journal with international circulation, but it caused no particular stir. An important reason is that they did not isolate the compound or compounds responsible and show that this or these were present in the suspect corn — presumably there was no chemist on the team capable of or interested in doing this; one of the sometimes unfortunate facts of research life is that most researchers are mainly interested in marching to the beat of their own drum, not to the beat of somebody else's drum. Had they isolated this compound and then found it in field material, they might have got the aflatoxin bandwagon rolling several years before it did. However, they did not.

The story of aflatoxin with all its multitudinous ramifications would fill, if not a library, at least a fairly long bookshelf all by itself, and the remainder of this chapter will be devoted to some aspects of it.

Identification of Aflatoxin and Aflatoxicoses

In 1960, 100,000 turkey poults and a lesser number of other kinds of domestic birds died in England, with a loss of at least several hundred thousand dollars. At first the deaths were tentatively attributed to a virus disease and the syndrome was named "turkey-X disease," the X of course denoting unknown cause. Some pretty sharp detective work on the part of those investigating the disease — a loss of several hundred thousand dollars was serious enough to stimulate research support — soon traced the cause to a batch of feed produced by Oil Cake Mills, Ltd., one of the largest manufacturers of oil cake feeds in Great Britain. Further work showed that the most probable toxic ingredient of this particular batch of feed was peanut meal. It was unlikely that the peanut meal itself was toxic, since peanut meal had long been used as a feed ingredient and was known to be an excellent source of protein; it must be, they reasoned, something added in some way to the peanuts or to the meal. One possibility was that the peanuts had been made toxic by toxin-producing fungi growing in them, and indeed isolations from the meal showed it to be heavily contaminated by *Aspergillus flavus*. The fungus was isolated, grown in pure culture, and fed to turkey poults. It killed them, with external signs and internal lesions identical to those detected in the birds that previously had died in the field.

Chemists soon isolated and identified the toxin and named it aflatoxin, the "a" from *Aspergillus* and the "fla" from *flavus*. As it turned out, there were four primary aflatoxins, later named B-1, B-2, G-1, and G-2, from their blue and green fluorescence, respectively, on thin-layer chromatography plates. There were an equal number of secondary aflatoxins, to be mentioned later, and possibly other and quite different toxins. Feeding tests with rations containing pure aflatoxin produced by the fungus grown in the laboratory very soon established the fact that, for those animals sensitive to it (all kinds of animals so far tested are sensitive to poisoning by aflatoxin, but to different degrees), extremely small amounts of the toxin consumed in food cause damage to various internal organs, especially to the liver. One of its effects on the liver is to induce the development of cancer.

Naturally this caused quite a furor, especially among the nutritionists and those concerned with problems of public health such as the National Institutes of Health and the Food and Drug Administration. Many questions were raised immediately, some of them frightening. Was the occurrence of aflatoxin in the particular batch of peanut meal incriminated as the source of aflatoxin in the turkey feed just a chance thing, or did many batches of peanuts contain aflatoxin? Ordinarily only the poorer grades of peanuts were processed for feed, the major portion of the crop being consumed by people, as peanuts themselves, raw or roasted, as peanut butter, and as peanut flour (some not-so-good grades of peanuts went into flour, too, since who could tell the difference?). Right about the time that aflatoxin was discovered and its dreadful effects on animals that consumed it began to become known, nutritionists in UNICEF as well as others were pushing peanuts as an excellent source of protein for infants in tropical countries where the diets often are deficient in protein, a deficiency resulting in widespread development of a disease known as ''kwashiorkor,'' the aftereffects of which include permanent physical stunting and mental retardation. Various low-priced rations, mostly in powder form, and usually consisting in part of peanut meal, were developed for general distribution in some of the tropical and subtropical countries. Were all the people who consumed peanuts or peanut products in any form exposing themselves to the danger of liver damage, liver cancer, and associated ills? Was a ration designed to be especially nourishing and wholesome actually or potentially or occasionally toxic? The fungus *Aspergillus flavus* was known to occur commonly in many kinds of plant materials, including stored grains — in other words, in the raw materials of many foods. The fungus also has long been used in the Orient to prepare various kinds of vegetable cheeses and sauces, including soysauce, and it even is used in the manufacture of diastase which in turn is used in baking and brewing in the United States. Did it produce aflatoxins in all of these things and, if so, under what conditions? Once the toxin was formed, did it persist through cooking or other processing? If so, could it be inactivated or in any way rendered innocuous? These and a multitude of other questions demanded almost immediate answers, answers that could be got only by research, and so the aflatoxin bandwagon got off to a roaring start.

One measure of the research effort devoted to a given problem is the number of research papers published on it. By this measure, the record for aflatoxin is impressive. In 1960, 16 research papers were published on aflatoxin; by 1966 the annual output had risen to 216 papers, and since then has settled to a fairly steady rate of close to 150 papers a year. A bibliography in 1970 listed 1200 papers on aflatoxin. In the ten-year period from 1962 on, 20 summary or review papers were published on aflatoxins specifically or on mycotoxins in general. It is fair to say that in the 13 years since aflatoxin was first isolated and recognized as a toxic metabolite, more research has been devoted to it than has been devoted to any other fungus-related problem of any sort in an equal period of time. And by its very nature much of this research is costly — it just is not something that a brilliant young investigator working out in the toolshed behind the house can contribute much to. Not all the answers on aflatoxins and aflatoxicoses are in yet, naturally, but at least we know a lot more about it than we did in 1960, and the following is an attempt to summarize some of the more important and interesting aspects of the problem. Those interested in pursuing this subject further are referred to the excellent book on aflatoxin edited by Goldblatt (32).

The Fungus

Aspergillus flavus is a "group species." The latter term is in quotation marks because in formal taxonomy such a taxon or taxonomic pigeonhole does not exist, and some of the pure (in a taxonomic sense) classificationists grind their teeth over the use of the term. However, the concept is biologically sound and useful, and it might well be applied to many other groups of living things; but, the nature of taxonomists being what it is, it probably will not be so applied — some taxonomists still labor under the doctrine of special creation, either by the Lord or by themselves, and they do not always make a clear-cut distinction between the two. The group is also known as *Aspergillus flavus-oryzae*. Within the group species there is also an individual species named *Aspergillus flavus*, and one known as *Aspergillus oryzae*; this might be supposed to lead to some confusion, but actually it does not especially.

According to Raper and Fennell (36), world-recognized authorities on this group, it includes eleven species, some of which grade into one another, which is the rule in any group of closely related species — and is what evolution is all about. Here, it or they will be referred to simply as *Aspergillus flavus.* To compound the confusion a bit, after aflatoxin was named it was found that *Aspergillus parasiticus* Speare, within the *A. flavus* group, was a more common and usually more potent producer of aflatoxin than was *A. flavus* Link (Speare and Link being the taxonomists who named these species).

These fungi are common and widespread in nature — in soil, in decaying vegetation, in hay and grains undergoing microbiological deterioration. *Aspergillus flavus* growing in these things can heat them up to 45°–50°C (113°–122°F) and hold the temperature there until the substance on which it is growing is just about totally consumed. It occurs in insects, and sometimes is parasitic in bees and flies; it grows on meat and meat products; some strains of it have been used since ancient times in Japan to saccharify rice for fermentation into sake; as already indicated, it is widely used to prepare vegetable cheeses and various kinds of sauces. *A. flavus*, in other words, is an aggressive and successful organism that does not lead a precarious existence in some narrow out-of-the-way ecological niche but abounds throughout the world, is present just about everywhere just about all the time, and is ready to invade all sorts of organic things whenever and wherever the conditions are favorable for its growth — mainly a high enough moisture content and a high enough temperature. Once it is established in a mass of stored grain or hay, it can for some time maintain the conditions that allow it to continue to grow, and often to predominate, in competition with all the other fungi likely to be present.

The conditions necessary for *Aspergillus flavus* to produce appreciable amounts of toxin are somewhat narrower and more restricted than are the conditions for it to grow — fortunately, since otherwise aflatoxin would be more generally present and perhaps a much greater hazard than it is. The major environmental factors that determine whether it will grow sufficiently to produce toxins are discussed below.

Environmental Factors in Producing Toxins

Moisture. To grow in any material, regardless of what the material may be, *Aspergillus flavus* needs a moisture content in that material in equilibrium with a relative humidity of 85 percent. This is the *minimum* moisture content at which it will grow. (Unlike some other fungi, *A. flavus* does not have a maximum moisture content for growth — the higher the moisture content above the lower limit, the faster the fungus grows.) In the starchy cereal seeds such as wheat, corn, rice, and sorghum, this is a moisture content of 18.5–19.0 percent, wet weight or as is basis. (Twenty percent wet weight basis is 25 percent dry weight basis. In grains and seeds the moisture is expressed on a wet weight basis because the water in the materials is bought, sold, and transported along with the other substances; it probably would be more sensible, and avoid a lot of grading and marketing problems, if moisture could be expressed on a dry weight basis, but this is not likely to come about before the arrival of the millennium.) In soybeans it is a moisture content of 17.0–17.5 percent, and in peanut seeds — the portion we eat — it is a moisture content of about 9.0 percent. At moisture contents below those in equilibrium with a relative humidity of 85 percent no material of any kind can be invaded by *A. flavus*. The importance of this will be obvious in a moment.

Aspergillus flavus is a member of a group known as "storage fungi" or "storage molds" — fungi that invade materials in storage. The term was coined in work with stored cereal grains and aimed to recognize and point out the fact that these storage fungi do not invade grains and seeds in the field, but only after they are harvested and stored. This group of storage fungi includes some half a dozen species of *Aspergillus*, plus a couple of species of *Penicillium* and *Sporendonema sebi*, and a few yeasts. All these fungi have the ability to grow without free water, and some of them not only endure but *require* a high osmotic pressure to grow; one of them will grow in agar that is saturated with sodium chloride, which no other organism on earth can do. None of these storage fungi invade grains or seeds to any extent before harvest — at high moisture contents they cannot compete with the many kinds of field fungi that always are present. Inoculum of these storage fungi, in the form of spores, probably is more or less ever present, but it is much more abundant in and near storage warehouses and within homes than it is out-of-doors. *A. flavus* does sometimes invade ears of corn in the field, when they have been attacked by earworms or, farther south, by weevils or other stored-products insects. And when it does so invade ears of corn in the field it may produce aflatoxin — but, in the United States only a few minor cases of such field infection have been found, and these were outside the major corn-growing area. Peanut fruits and seeds rarely are infected by *A. flavus* before harvest; their period of high infection hazard is between the time the vines, with the pods attached, are pulled from the ground and piled in windrows, and the time the pods are removed from the vines. Cotton bolls and the seeds within them sometimes are infected with *A. flavus*, but only in two relatively small areas in the United States, and then only when the weather at harvest time or just before is unusually humid; the time of major production of aflatoxin in cottonseed is during storage.

Temperature. The minimum temperature for the production of aflatoxin by *Aspergillus flavus* is 12°C, the optimum is 27°C, and the maximum is 42°C (54°, 81°, and 108°F, respectively). The fungus will grow slowly below the minimum temperature for the production of aflatoxin, and will grow rapidly at a temperature 5° to 10°C (9° to 18°F) above the maximum temperature at which aflatoxin is produced. When *A. flavus* grows as a predominant organism in stored corn, rice, or other grain it may raise the temperature of the grain in which it is growing to 45°–50°C (113°–122°F) and hold it there for some time; such grain may be caked and yellow-green with spores of *A. flavus*, and one would expect it to be loaded with aflatoxin, whereas in fact it contains little or none — either the temperature was too high for aflatoxin formation or the presence of other organisms prevented it or both.

Presence of other microflora. *Aspergillus flavus* seldom invades stored grains alone, as a pure culture; if the grains are moist enough to permit invasion by fungi, they almost always are invaded first by *A. glaucus*, followed by other species, especially *A. candidus* and the yeastlike *Candida pseudotropicalis* and then, when as a result of the metabolic activity of these fungi the moisture content has been raised to over 18 percent and the temperature to 35° or 40°C (95°–104°F), *A. flavus* may take over. The exact makeup of the fungus flora will not always be the same, and neither will the sequence always be the same as that given, but the general pattern is that *A. flavus*, when growing in grains and seeds stored in bins or barges or what have you, seldom grows alone but is preceded, accompanied, and followed by other microflora, principally fungi. (See plate 3.) There has been little work on aflatoxin production or lack of it in such mixed cultures, since relatively few of those who have investigated the formation of aflatoxin have been interested in the conditions under which it is *not* formed. However, such work as has been done indicates that if the fungi normally present on grains are allowed to grow, then even if the grain is inoculated heavily with a known aflatoxin-producing strain of *A. flavus*, and the fungus is allowed to grow for some time, no aflatoxin will be formed.

In our mycotoxin work at the University of Minnesota we have, over the past several years, made repeated tests in which grain was conditioned to a moisture content of 22–25 percent, wet weight basis (high enough for *Aspergillus flavus* to grow vigorously), was inoculated heavily with a known aflatoxin-producing strain of *A. flavus*, then was held for 10 to 14 days at a temperature optimal for aflatoxin production. (See plate 4.) The fungi normally present were allowed to grow, as they would in grain in a bin. The grain, by that time caked and bound together by fungi and often yellow with the spores of *A. flavus*, was dried, ground, and incorporated into a balanced ration as anywhere from 10 to 50 percent of the total and fed to various kinds of experimental animals — ducklings, which are very sensitive to injury by aflatoxin, white rats, baby chicks. Some of these feeding tests were continued for eight months. In no single case were the animals that were given the ration containing the heavily molded grain — the molds including a potentially toxic strain of *A. flavus* — deleteriously affected in any way. Their weight gain was the same as that of the animals on control feed containing food-grade grain, their feed efficiency was the same or, in some cases, slightly higher than that of the animals on the control feed, and when the animals were sacrificed and necropsied at the end of the tests no lesions or abnormalities were detected. A number of authorities in high places have at times made some rather strong statements to the effect that *any* feed moderately to heavily invaded by fungi probably is harmful. There actually has been no experimental evidence to support this, and our evidence, although limited, indicates that it just is not true. Also growers have sued feed manufacturers for sometimes almost astronomical amounts of money claiming that a given batch of feed contained molds that caused injury or death in their cattle. What an adroit lawyer can make a hometown jury believe is one thing, but the evidence to date is that feed even heavily invaded by a mixture of molds, including some known to produce toxins, is not necessarily at all harmful. Evidently these toxin-producing fungi must occur alone, essentially as pure cultures, to produce appreciable amounts of toxin. Some of the high-aflatoxin-risk materials are precisely those in which *A. flavus* is likely to occur alone.

Substrate. Aflatoxin has been found in many raw and manufactured products in nature — in the field, warehouse, or home — and has been produced in many more kinds of materials in the laboratory. A 1966 summary of aflatoxins in foods listed the following in which aflatoxins (the toxins, not just the fungus) had been found up to that time: cassava, cocoa, coconut, corn, cottonseed meal, fishmeal, peanuts and peanut meal, peas, potatoes, rice, sake, soybeans, and wheat. *Aspergillus flavus* has been found in country cured hams and in sausages; when such hams were inoculated with *A. flavus* in the laboratory and held for a time, the fungus grew in them and produced aflatoxin, which shows that aflatoxin *can* be produced in hams, but not that it *is* produced in them as they are cured in practice. When *A. flavus* is growing in these hams as they cure or age in the curing rooms, it must almost inevitably be accompanied by a multitude of other organisms which, according to the account given above concerning mixed cultures, makes for low aflatoxin risk.

Given an equal growth of *Aspergillus flavus*, much more aflatoxin will be formed in one substrate than in another. The fungus, for example, grows about equally well in both soybeans and peanuts, yet very little aflatoxin is produced in soybeans, whereas large, sometimes fantastically large, amounts may be produced in peanuts. Thus some food and feed materials can be rated as of low or moderate aflatoxin risk, and others as of high aflatoxin risk. The degree of aflatoxin risk is affected by weather and climate, too, since in areas of high temperature and humidity everything gets moldier than it does in areas of dry air and low temperature. The way foods and feeds, or foodstuffs and feedstuffs, are handled also affects aflatoxin risk; the Bantu tribes in Africa *prefer* the sour flavor of partly spoiled corn to that of sound corn; they also (coincidentally or not) have a very high incidence of primary liver cancer. In some regions where refrigeration is not available it is customary to cook up enough rice, or whatever else constitutes the staple diet, to last for several days, so that a portion of the batch may be exposed for two or three days to invasion by fungi; if the fungus is a toxin-producing strain of *A. flavus* and the region is hot and humid, appreciable amounts of aflatoxin may be produced in that particular batch of food before it is consumed. A combination of high-aflatoxin-risk material, high-aflatoxin-risk climate, and high-aflatoxin-risk practices in handling foods of course increases the chances of aflatoxin poisoning.

Peanuts, also known as earthnuts and groundnuts, probably head the list of high-aflatoxin-risk foods that are consumed in quantity by a large proportion of the population in many countries. Up until a few years ago Brazil nuts probably contained as much aflatoxin, on the average, as did peanuts, but they never have formed a regular portion of the diet of any people; some other kinds of nuts evidently furnish about as good a substrate for aflatoxin production as do peanuts, but they seldom are consumed in quantity.

The peanut plant, *Arachis hypogea*, is a native of subtropical America, as are some 20 or more other kinds of economic plants that now are cultivated around the world. Soon after the flowers of the peanut plant are fertilized, the tips of the stems on which the flowers bloom turn downward and penetrate into the earth; the pods, or fruits, are borne on these underground pegs. At harvest the whole plants are pulled up and piled in windrows or in loose stacks until the pods become dry enough to be easily removed from the vines. *Aspergillus flavus* invades the seeds within the pods during this drying period. If the weather at harvest is humid, as it often is in many countries where peanuts are grown, the risk of infection by *A. flavus*, and of aflatoxin production, is very high. This infection and aflatoxin production may or may not be accompanied by obvious discoloration of the pods or of the seeds within them, but it is likely to be accompanied by sporulation of the fungus either on the outside of or within the pods or shells. This can be detected by close inspection conducted by someone who can recognize the fungus. In the United States representative samples are taken from each truckload as the peanuts come to market and, if any of those inspected are found to harbor the fungus, that lot is diverted for processing into feed or fertilizer.

As soon as the aflatoxin problem surfaced in the early 1960s, programs were established by the peanut growers themselves, aided by concentrated research, to reduce the possibility of aflatoxin occurring in edible peanuts and peanut products. That these have been successful is indicated by the fact that although random sampling by the Food and Drug Administration laboratories may turn up an occasional batch of peanut butter that contains aflatoxin, it almost never is found in sufficient amount to be of any real concern. Before 1965 anyone who consumed peanuts in any form in the United States, and especially those who consumed products made in part from peanut flour, must at times have consumed an appreciable amount of aflatoxin too. In many countries with no pure food regulatory agencies, and where peanuts furnish almost the only good source of protein, the people must continue to consume appreciable amounts of aflatoxin. So must some of the health food addicts who get their peanuts directly from the farm, and those who eat the peanuts that they themselves grow.

Peanut meal and peanut cake, from which the oil has been removed, have long been used as high-protein ingredients in animal feeds. In northern Europe, where animal industry is so important a part of the economy, these peanut products have to be imported, principally from the United States, Brazil, Central Africa, and India. In Denmark, where 52 samples of peanuts and peanut products (whole nuts, meal, and cake) imported for feed from ten countries were tested, aflatoxin was found in 86.5 percent of them, and one sample contained 3465 micrograms of aflatoxin per kilogram of the peanut meal. Expressed another way, this was 3465 parts of aflatoxin per billion parts of the product, usually expressed as ppb. As a result of that the Danish Ministry of Agriculture established a limit of 100 ppb of aflatoxin in peanut products imported for feed. This is five times as high as the 20 ppb of aflatoxin permitted in feedstuffs in the United States, according to Food and Drug Administration regulations — but the establishment by the Danes of this limit of 100 ppb probably was a practical recognition of the fact that most peanut-exporting countries could not meet a lower limit. Thus batches of peanuts or of peanut products that are not permitted to be sold as feed in the United States because they have more than 20 ppb of aflatoxin can be sold to other countries where the standards are less strict. This has been deplored by some; however, it is all perfectly aboveboard, and any country is free to establish such "tolerance limits" as it chooses. Also, the peanuts in animal rations make up only a portion of the total ingredients, probably seldom as much as 25 percent. Assuming that the other ingredients in the ration are free of aflatoxin, this means that the feed consumed by the animals contains, at most, 25 ppb of aflatoxin. Feeding tests referred to below with rations containing ten times this amount of aflatoxin have shown no ill effects of any kind in swine or cattle. However, anyone who travels or sojourns in tropical countries *anywhere* in the world would do well not to eat peanuts or peanut products in any form, *ever*. So much for peanuts.

Seeds of cereal crops — wheat, corn, barley, oats, and sorghum — and those of soybeans are, in general, of low-aflatoxin-risk. At the Northern Regional Research Laboratory, in Peoria, Illinois, more than 2000 samples of these grains were collected in the late 1960s and early 1970s from federal grain inspection offices in the major grain marketing centers in the United States and tested for aflatoxin. Aflatoxin in small amounts was found in only a very few samples, and those in the lowest grades of the various grains. The researchers stated (39), "Our results indicate that the factors used to grade these samples would probably exclude from the food markets samples likely to contain aflatoxin." And, "according to recent publications, the levels at which we detected aflatoxin in corn and soybeans would not be injurious to swine and cattle." The same might not be true of, say, grain grown and fed on the farm, or traded locally without more than casual and perhaps inexpert inspection, or in countries where there are no formal grades and no formal inspection of grains in commerce to help ensure high quality. Our mycotoxin laboratory at the University of Minnesota, for example, has received corn from a food market near New Delhi, India, that had a high percentage of the kernels invaded by *Aspergillus flavus* (although the kernels did not disclose this by any outward sign) and that contained about 80 ppb of aflatoxin. Corn grown and stored in the tropics may be of higher aflatoxin risk than is corn grown and stored in the corn belt of the United States.

Time. Other things being at the optimum for aflatoxin production — moisture content, temperature, and substrate plus the presence of a toxin-producing strain of *Aspergillus flavus* alone — some aflatoxin can be produced within 24 hours. A maximum will be reached in ten days or two weeks, after which the amount present may decrease rapidly.

Strains of *Aspergillus Flavus*

When aflatoxin first came into prominence some people supposed that aflatoxin most likely was produced by only odd and unusual strains of the fungus — after all, *Aspergillus flavus* had long been used to prepare various kinds of food products that, to all appearances, were perfectly wholesome, and so the assumption that toxin-producing strains of the fungus were uncommon seemed reasonable. Reasonable, perhaps, but untrue. Workers in Texas collected 284 strains of *A. flavus*, most or all of them from rice, and tested them in the laboratory for their ability to produce aflatoxin. Of these 284 isolates, 268, or about 94 percent, produced some aflatoxin, and 86 isolates, or 33 percent of the total, produced more than 10,000 ppb, and 4 isolates produced more than 250,000 ppb of aflatoxin. Granted, this was in laboratory tests under conditions supposedly highly favorable to the production of aflatoxin; even so, there is no question but what strains of the fungus capable of producing large amounts of aflatoxin are common — anyone who in his work with aflatoxin has examined different isolates for their ability to produce aflatoxin has found such strains in abundance. At the Northern Regional Research Laboratory toxin-producing strains of *A.*

flavus were obtained from peanuts from just about all over the world where peanuts are grown, from rice, wheat, corn, cottonseed hulls, soil, sugarcane stalks, mealy bugs, and diseased bees. At Minnesota we have found them to be common in whole and ground black pepper, in ground red pepper, and in macaroni and spaghetti and other pasta products. These latter food products warrant a few words.

Pepper. Black pepper is the fruit of *Piper nigrum* and is produced chiefly in India and Indonesia; white pepper consists of the seeds of the same plant but divested of the outer tissues (they used to do this by soaking the black pepper corns or fruits in water in vats for a time, then trampling it out by foot to remove the outer pulp — ugh!). These spices do not undergo any processing, other than drying and grinding, before being added to foods.

Over a period of several years approximately 100 samples of whole and ground black pepper were examined in our laboratory for number and kinds of fungi — all the samples came from local retail or wholesale stores or from the small portions served on airplanes. One of the first samples of whole peppercorns that was examined — it happened to be a two-ounce can bought in a neighborhood store — contained a rodent dropping of about the same size as the peppercorns, doubtless a southeast Asia rodent. What a gourmet's delight! Most samples of whole peppercorns contained some to many peppercorns partly eaten by insects and partly or mostly decayed by fungi and bacteria. In dilution cultures, the number of fungus colonies in whole or ground pepper averaged 52,000 per gram and ranged up to over half a million per gram; they consisted principally of *Aspergillus flavus*, *A. ochraceus*, and *A. versicolor*, all three of which are capable of producing potent toxins. Some of the samples of ground pepper bought in one-pound or five-pound tins were lightly caked with fungus mycelium when first opened in the laboratory, and, with time, a number of these became solidly caked with fungi and covered with a felt of mycelium. (See figures 3-1 and 3-2 and plate 1.)

The figures above may not give much idea of the degree of contamination by fungi of these samples of pepper, so a rough comparison may be helpful. Wheat intended for milling into flour seldom contains more than a few thousand colonies of storage fungi per gram of grain, and

Figure 3-1. A rodent dropping (top) and a small stone from a sample of "pure" black peppercorns

neither does the flour milled from the wheat. If a sample of flour intended for baking was found to have even one tenth as many colonies of fungi and bacteria per gram as many of these samples of "pure" black pepper had, the regulatory agencies would come down vigorously on the miller. If barley has as many as 10,000 colonies of some of the

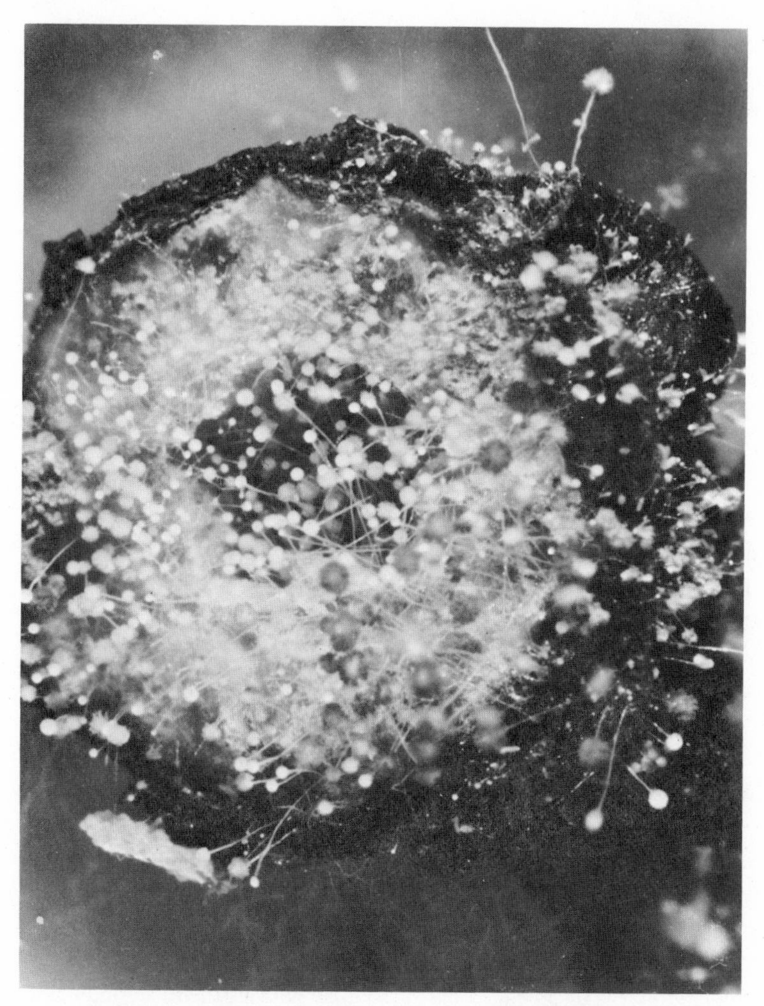

Figure 3-2. A ''pure'' black peppercorn cut in half and placed on agar. *Aspergillus flavus* and *A. ochraceus* are growing out from the entire interior.

same kinds of fungi per gram as were found in these samples of ''pure'' black pepper, it would be rejected for malting. If bland foods such as flour or breakfast cereals or bread were as heavily invaded by fungi and bacteria as were most of the samples of ''pure'' black pepper tested, they would have so musty an odor and taste that they would be too revolting to eat — the natural spicy odor and flavor of black and white pepper are potent enough to conceal any odor of decay. In a way it is ironic that black pepper, which supposedly first came into use, a few hundred years ago, to help conceal the odor of partly spoiled meat, should itself be spoiled.

Only a relatively few samples of black pepper were tested for bacteria, but in these few the bacteria averaged 200,000,000 per gram of pepper, and the count ranged up to nearly a billion, a really astronomical figure. Among the bacteria isolated were *Escherischia coli*, the common fecal bacterium, and *Staphylococcus* and *Streptococcus*, some species of which cause infectious disease in man.

Some of the isolates of *Aspergillus flavus* from black pepper were grown in autoclaved moist rice, which then was dried and incorporated into a ration and fed to ducklings, and it killed them within a few days. Attempts to detect aflatoxin in the samples of pepper that yielded large numbers of colonies of *A. flavus* were unsuccessful because of the difficulty in getting rid of the ''garbage'' or extraneous materials extracted from the pepper by the solvents; to overcome that problem would have required a research project in itself. However, extracts from a number of samples of black pepper as they came from the stores were combined, condensed, and added to a ration given to rats, and all the rats died within two to four days. No one has thought to question whether black pepper itself might be toxic. Most of us, of course, consume a relatively small quantity of black pepper; a modest single serving, or sprinkling, of black pepper amounts to about 50 milligrams; 100 milligrams per day over a period of 60 years adds up to 77.25 ounces or 4.82 pounds — counting, of course, the crud as well as the ''pure'' black pepper. Some gourmet recipes specify that, for proper preparation, the dish be flavored at the table with a generous grinding of whole black pepper from a fancy pepper mill. Suppose that a dish of macaroni or rice is so flavored, and thereby is inoculated with a heavy load of fungi that is potentially toxin producing and of infectious bacteria, and then a portion of the dish is left over, unrefrigerated, for a day or two; it might still be edible and even wholesome, but to anyone who is familiar with the microbiology of ''pure'' black pepper it could hardly be attractive.

The January 5, 1972, issue of the *Wall Street Journal* carried a front-page article intriguingly entitled "Does Your Paprika Get up Off the Plate and Just Walk Away?" The story dealt with contamination of spices by living and dead insects, insect fragments, bacteria, fungi, and miscellaneous "extraneous material"; it included a brief presentation of points of view of members of the American Spice Trade Association (who evidently believe that no problem exists, since the president of the association was quoted as saying, "Insect fragments won't make you sick — in fact, they're protein" — what a cavalier and callous attitude toward the public which consumes this expensive filth!); of officials in the FDA (who evidently were somewhat concerned about filth in spices but who were not doing much about it); and by representatives of the Consumers Union (who think that the "pure" on the label of a container of "pure black pepper" should mean what it says and that rigorous standards for the cleanliness of spices should be established and enforced). That portion of the article that dealt with microbial contamination of black and white pepper was based on the work done at Minnesota, described in a 1967 publication (31), which was later supplemented by a couple of additional papers on the same subject (30, 35).

The article quoted men in the spice trade as saying that, to have clean pepper, someone would have to go over the peppercorns one by one and pick out the bad ones with tweezers. This is nonsense. First, no one can recognize "bad" peppercorns with the bare eye, or even, in most cases, with a microscope — by the time any food product is so heavily invaded by fungi and bacteria as to be obviously decayed, it is in the last stages of decay. Many of the peppercorns that we examined with a microscope appeared sound outwardly but when sectioned were found to be totally decayed within. Second, the peppercorns come from the vines clean and sound, and if cared for properly during storage and shipment they should remain clean and sound until they reach the consumer. As a comparison, malting barley is harvested, transported, stored, and processed in amounts of tens of millions of bushels per year in the United States. Most of it comes to market with close to 0 percent of the kernels invaded by storage fungi. It may be stored for a year or more, and after that it still is free of storage fungi. It also is very low in insects, insect fragments, and other ''extraneous'' material. This happy state of affairs did not come about by chance; it is the result of practices established on a firm base of knowledge by microbiologically educated and knowing men in the malting industry. Black pepper, by the time it reaches the consumer, costs about 100 times as much as does malting barley, and it could easily be well cared for — all that is necessary is to dry it to a moisture content of 13.0 percent or below right after harvest, keep it below that moisture content throughout its storage life, and store it in clean containers in rat- and insect-free surroundings.

Pasta products. Macaroni, spaghetti, and other pasta products are made from semolina, a coarse-ground flour from durum wheat. The process is simple: semolina and water are mixed to a thick dough in a mechanical mixer, this dough is forced through dies into whatever forms are desired, and these are dried in cabinets or on racks and packed for sale. The drying goes on slowly over a period of several days — fast drying results in case-hardened products that are unattractive to customers. During this drying, especially in warm, humid weather, the products may be invaded by bacteria, yeasts, and filamentous fungi. In summer in Kansas City, for example, you can locate a macaroni factory a considerable distance downwind by the aroma, and it is not the aroma of durum flour.

The kinds of fungi present in about 50 samples of pasta products were determined by plating pieces of them on agar media favorable for the growth of various fungi. The samples came from factories in 12 cities of the United States and one in Canada, and all were bought in stores and supermarkets in and near St. Paul, Minnesota, over a two-year period. All of these samples yielded a rich and varied fungus flora; in nearly all the predominant fungus was *Aspergillus flavus*, followed by *Aspergillus ochraceus* and *Penicillium*. (See plate 2.) Some of the isolates of *A. flavus* from these pasta products were grown in moist autoclaved corn for ten days; then the corn was dried, ground, mixed into an otherwise balanced ration, and fed to ducklings. Several of the isolates so fed resulted in the death of the ducklings in a few days, and the others resulted in greatly reduced weight gain. The probability is that some and perhaps many of the isolates of *A. flavus* from pasta products are capable, when growing alone under favorable circumstances, of producing aflatoxin. Van Walbeek *et al.* (41) reported 12.5 ppb of aflatoxin in a sample of spaghetti that was suspected of having caused illness in children who had eaten it. We found no aflatoxin in the samples we tested for it. However, why the rich and varied population of microflora in these products? These fungi certainly do not come from the semolina, since we have tested many samples of that and they uniformly are almost free of microflora. Almost certainly the microflora get into the products during or immediately after mixing, and they grow during drying. It should be relatively easy to ferret out the source or sources of fungus contamination in macaroni and spaghetti factories and to eliminate them simply by good housekeeping. Some of the microflora present in such large numbers in these products may be desirable in that they possibly contribute to the delicious flavor, but it seems hardly likely that potential aflatoxin-producing strains of *A. flavus* would be included among these.

Foods processed with the aid of Aspergillus flavus. As mentioned above, soysauce is prepared with the aid of *Aspergillus flavus*, and so are a number of other sauces widely used in the Orient, as well as the so-called vegetable cheeses. Where these products are manufactured on a fairly large scale, stock cultures of the suitable strains of the fungus are maintained in one way or another, and these are added to the other ingredients. Many strains of *A. flavus* used in these various processes have been tested, and none have been found to produce aflatoxin. However, where such products are prepared as a household or village industry, and the strain of the fungus is just carried along from one batch to the next, as yeasts or as sourdough bacteria are carried along in home baking, wild strains capable of producing aflatoxin may contaminate them. In a recent investigation of a sauce so prepared and widely used by families in the Philippines, not a single sample of it was found to be free of aflatoxin. In some of those communities probably everyone was suffering to some extent from chronic aflatoxin poisoning.

Effects of Aflatoxin on Animals

Different kinds of animals differ in sensitivity to damage by aflatoxin, as they differ in sensitivity to most other kinds of toxins. So far as can be seen there is no rhyme or reason to this differential sensitivity — it's just the way things are.

Rainbow trout are extremely sensitive to damage by aflatoxin, but some of their relatives are not particularly so. In one test, rainbow trout were given a ration containing 20 parts of aflatoxin per billion parts of feed; after a year on this ration, 39 of the 78 rainbow trout in the test had liver tumors, and another 26 had suspected tumors. Twenty parts per billion is equal to one pound of aflatoxin in 50,000,000 pounds of feed, or one ounce of aflatoxin in 3,125,800 pounds of feed, or one ounce in 1562 tons. The ordinary freight car used for hauling grain in bulk has a capacity of about 20 tons, so 1562 tons would fill 78 cars — a fairly long train; and through this rather formidable bulk would be distributed the one ounce of aflatoxin. Or it would be equal to one drop in 1320 gallons.

Ducklings also are injured by very small amounts of aflatoxin in their diet. One standard method of detecting aflatoxin in feeds and foods is to feed samples to ducklings for a time — if they become ill or die their livers are examined for the kind of cellular changes that are practically diagnostic for aflatoxin poisoning. Ducklings seven days old were given a ration containing 30 parts of aflatoxin per billion parts of feed, and after four weeks 16 of 37 birds in the test had died or had lost weight, and had developed the cancerous liver lesions characteristic of aflatoxin damage.

Rats are not particularly sensitive to damage by aflatoxin, but when white rats were given a ration containing only 15 parts of aflatoxin per billion parts of feed, nearly all of them eventually developed liver cancer. It is no wonder that aflatoxin has been said to be the most potent carcinogenic or cancer-causing substance known. It obviously is not something that one wants to consume, even occasionally and in small amounts, if it can be avoided. In some countries it probably cannot be avoided. People who work with pure aflatoxin in the laboratory are likely to take extreme precautions to avoid contamination of the premises or of themselves with aflatoxin, but even with the most stringent precautions the laboratory sometimes turns out to be generally contaminated.

In domestic animals the susceptibility to injury by aflatoxin ingestion goes in about this order, from the most to the least sensitive: young pigs and pregnant sows, calves, fattening pigs, mature cattle, sheep. In all of these animals, consumption of aflatoxin in amounts too small to produce obvious symptoms may result in stunting or in lack of normal weight gain, in general unthriftiness or "poor doing," and in increased susceptibility to infectious diseases. Keyl *et al.* (33) fed different amounts of aflatoxin to swine for 120 days and to Hereford beef steers for four and a half months. In swine they found no evidence of any toxic effects at levels of 233 ppb dietary aflatoxin or below, and in beef cattle no evidence of any toxic effects at levels of 300 ppb dietary aflatoxin or below.

In the animals so far studied, most of the aflatoxin ingested is taken up by the liver, some by other organs, and some of it is excreted in the urine and feces and, in lactating mammals, in altered but still toxic form in the milk. Brewington and Weirach (28) tested many hundreds of samples of fluid and dried milk from different marketing areas in the United States and found no aflatoxin in any of them. The possibility of anyone in the United States getting in dairy products enough aflatoxin to be injurious seems to be pretty remote.

Aflatoxin in Human Health

Where man rates in sensitivity to damage by aflatoxin is not known, since there have been no feeding tests to determine this. There is, however, an abundance of circumstantial evidence, plus some fairly direct evidence, that ingestion of aflatoxin-contaminated food results in the same sort of damage to man as it does in other animals. Primary hepatocarcinomas — cancers which originate in the liver — are much more prevalent in some countries than in others (34); the incidence of such cancers is more than 100 times as high in some countries of Africa and Southeast Asia as in some of the northern European countries — and it is in precisely those countries of high incidence of primary hepatocarcinomas that aflatoxin is likely to be consumed in the diet. In some portions of India and Southeast Asia primary hepatocarcinomas occur even in nursing children, and the evidence is that this probably is due to aflatoxin poisoning. The mother consumes maize, rice, peanuts, millet, or some other product that contains aflatoxin. Enough of this aflatoxin will be excreted in the mother's milk to engender cancer in the infant. In Thailand, aflatoxin was found in many foods that were consumed regularly, especially by the rural population (37). The incidence of liver cancer in different provinces there was correlated with the dietary intake of aflatoxin. In some cases of fatal illness of previously unknown etiology, aflatoxin was detected in the various organs of the victims in sufficient amounts to establish beyond any reasonable doubt that it was the true cause of the illness and death. The same situation must prevail in many areas of the warm and humid tropics around the world.

Detection of Aflatoxin

To establish limits for the amount of aflatoxin permitted in foods or feeds presupposes the existence of methods that enable one to extract the toxin from different materials, to purify it, and to measure the amount present with fairly small limits of error and a fairly high degree of consistency. A good deal of very expert research has been devoted to this aspect of the problem. Basically, to determine whether aflatoxin is present and, if so, how much is present, a sample of about 50 grams of the test material is extracted with a special solvent or solvents, the extract is condensed to a small amount, this is put on a thin-layer chromatograph plate and allowed to develop for a time, after which the plate is examined under ultraviolet light. A bright bluish spot at a given location on the plate is *indicative* of aflatoxin — but indicative only. The material of the spot is scraped off, taken up in a solvent, and subjected to analysis by ultraviolet spectrography. If the absorption bands here indicate aflatoxin also, this is just about conclusive, but sometimes it is desirable to confirm the finding by adding some of this material to an otherwise pure ration which then is fed to day-old ducklings. If they become ill or die, and microscopic examination of sections of their livers reveals damage typical of that caused by aflatoxin, and almost only by aflatoxin, the diagnosis is about as certain as it can be. The "otherwise pure ration" mentioned above is not just any commerical poultry or duckling ration — they sometimes contain aflatoxin — but a ration called a "toxicological diet" that is compounded of materials of known purity.

The amount of aflatoxin present is determined by comparing the brightness of the fluorescent spot on the chromatographic plate with a standard made up with pure aflatoxin, and so it is to some extent subjective. Samples containing known amounts of aflatoxin were sent to several laboratories where work on aflatoxin detection was going on, and they were asked to report the amount of aflatoxin present. On one sample one laboratory reported 6.5 ppb, and on the same sample another laboratory reported "between 10 and 50 ppb." No one can with confidence say that a sample of naturally contaminated material contains exactly 6.5 ppb of aflatoxin — the laboratory workers who reported that just were not aware of their own limitations. If a limit of 20 ppb is going to be established as the amount permissible in feeds, the methods had better be good enough so that the ordinary quality-control laboratory can use them to distinguish between 15 ppb, which is permitted, and 25 ppb, which means condemnation.

The procedures that serve to extract aflatoxin from one material may not work for another material. Then too any solvent that extracts the aflatoxin will extract many other things, and these extraneous materials (referred to as "garbage") must be separated from the aflatoxin. To develop a method to detect aflatoxin in a given material may require a separate research project.

For those who store, ship, and process grains such as corn, some sort of quick test for the detection of aflatoxin is essential, and a procedure has been developed that appears to be at least moderately reliable. If aflatoxin is present in a given lot of corn, it is likely to be present in the damaged kernels, and in what is called the "foreign material," which by definition is anything that will pass through a 12/64-inch round-hole sieve (including pieces or fragments of corn kernels). A representative sample is taken and screened to get the foreign material; this is then examined under ultraviolet light. If aflatoxin is present in a given fragment, that fragment will fluoresce a bright yellow-green. If there is no fluorescence, the sample is pretty certain to be free of aflatoxin. To confirm this, the particles that fluoresce can be extracted quickly with a solvent and this substance can be developed on a thin-layer chromatograph plate and examined under ultraviolet. This method of detection of aflatoxin by examination of corn kernels and portions of kernels with ultraviolet light works for corn, but not for other grains or for soybeans.

During the past few years several lots of corn in the United States have been found by the FDA to have high enough levels of aflatoxin to be condemned. News of this sort travels through the grain-merchandising and feed-producing industries like the shock wave of an earthquake. Justifiably so. Their survival depends on their being able to keep the feed ingredients and feeds free of aflatoxin and free of other mycotoxins for which low or zero tolerances have been established. Some of the larger firms have their own quality control laboratories where the rather involved analytical work to detect these toxins can be done. Small feed mills supposedly have available the services of testing laboratories for tasks such as this. From what has been said above, you can see that tests for the detection of aflatoxin require some rather expensive equipment and expertise, and a single test may cost $25 to $50. So if the feed mill manager wants 10 to 20 samples tested per day, he has an outlay of from $250 to $1000 per day for this alone. No small and few large feed mills anywhere in the world could survive under those conditions. They have a problem.

Control of Aflatoxin

Once aflatoxin is formed it is likely to persist; it is heat stable and is not destroyed by ordinary cooking; it will even endure autoclaving or pressure cooking for some time without much loss in potency. It also resists destruction or inactivation by most of the multitude of chemicals that have been tested for this purpose. Peanut meal containing aflatoxin has been detoxified by treatment with sodium hydroxide, ammonia, ozone, and methylamine, but none of these offer a commercially feasible treatment. Another approach has been experimenting with microorganisms that might convert the toxins into harmless compounds. About a thousand kinds of fungi, bacteria, yeasts, actinomycetes, and algae have been tested for this purpose, and one bacterium, *Flavobacterium aurantiacum*, detoxified milk, corn oil, peanut butter, peanuts, corn, and soybeans that had been artificially contaminated with aflatoxin — which may not be exactly the same as naturally contaminated materials in which the fungus has grown.

The only sound approach to the aflatoxin problem is prevention — to store, handle, ship, and process high-aflatoxin-hazard products so that they cannot become invaded by *Aspergillus flavus*. Hand in hand with this must go a relatively foolproof system of surveillance — sampling and testing to make sure that no products significantly contaminated with aflatoxin escape detection. In the developed countries this is, if not exactly easy, at least feasible; in the less developed countries it is not, and in some of those countries aflatoxin poisoning is likely to remain a public health problem for a long time to come.

4

Mycotoxins and Mycotoxicoses: Other *Aspergillus* Species, *Penicillium*, and *Fusarium*

Aspergillus Ochraceus and Ochratoxin

The Fungus

Aspergillus ochraceus, like *A. flavus*, is a group species that, according to Raper and Fennell (36), includes nine individual species. It is or these are common in soil, in decaying vegetation, and in stored seeds and grains undergoing microbiological deterioration. It can invade materials that have a moisture content in equilibrium with a relative humidity of 80 percent or above, which in the starchy cereal seeds means a moisture content of about 15.5 percent and above, wet weight basis. If wheat is conditioned to a moisture content of 17.0 percent, inoculated with spores of *A. ochraceus*, and held at a temperature of 25° C (77° F), within two weeks the germs or embryos of the kernals will be heavily invaded by the fungus, as shown in figure 4-1. Yet this fungus seldom is isolated from more than a small percentage of seeds or grains that are undergoing microbiological deterioration in bins or barges or ship holds, the reason evidently being that it is not a good competitor. When grains and seeds are stored under conditions that permit them to be invaded by storage fungi, they almost always are invaded first by the *A. glaucus* group, even if their moisture content is high enough to permit invasion by some other species of *Aspergillus* that require higher moisture contents than does *A. glaucus*. If inoculum of both *A. glaucus* and

Figure 4-1. *Aspergillus ochraceus* growing out from the germ or embryo of a wheat kernel

A. ochraceus is present on the seeds, as it almost always is, and if the moisture content is high enough so that *A. ochraceus* can grow, *A. glaucus* will predominate, sometimes to the almost total exclusion of *A. ochraceus*. This at least seems to be the general rule, but it may not hold for all times and places and substrates. In Mexico *A. ochraceus* sometimes has been isolated from 40 percent or more of surface-disinfected kernels of corn from bins in which deterioration was in progress. It also has been the major organism in some lots of whole black pepper, and dilution cultures of these samples made in our laboratory at the University of Minnesota yielded up to thirty or forty thousand colonies of *A. ochraceus* per gram of pepper, a far larger number of colonies of this fungus than we have ever encountered in wheat or corn even when these were badly deteriorated. These peppercorns were produced in the tropics, of course, since that is where pepper plants grow, but no one knows where the peppercorns were stored when they became invaded by this fungus and by associated storage fungi. We also found some samples of macaroni and spaghetti to be heavily invaded by *A.*

ochraceus, including those manufactured as far north as Minneapolis–St. Paul, Minnesota, so whatever the conditions are that make for heavy invasion of materials by *A. ochraceus*, they certainly do at times prevail far outside the tropics.

Toxin Production

Production of a toxin, named ochratoxin, by *Aspergillus ochraceus* was first described in South Africa (55). There was no implication at the time that the grain from which the fungus was isolated there was toxic; *A. ochraceus* was just one of a number of fungi isolated in more or less routine plating out of grain samples, and was grown in pure culture and tested for toxicity as a matter of interest, as were some of the others. The workers there also isolated, purified, and characterized the toxin chemically, a relatively formidable task.

The LD_{50} (the single dose that will kill 50 percent of the individual animals in a test) of ochratoxin for rats is 22 mg/kg, which equals 22 parts per million, but ingestion of lesser amounts will result in severe liver damage. A single dose of 12.5 mg/kg (12.5 milligrams of the toxin per kilogram of body weight of the rats) was administered to pregnant rats on the tenth day of gestation, and of the 88 fetuses involved, 72, or 81.8 percent, died or were resorbed. This experiment was made because death and resorption of fetuses, and abortion, are common in dairy cattle and swine, and usually cannot be attributed to any known cause; many cases of this nature are associated with consumption of hay or other feed heavily invaded by molds. As is indicated later, this does not mean very much, because many samples of feed are heavily molded or are made up of ingredients that have at some time been heavily invaded by fungi, including potentially toxic fungi, and yet are wholesome and nutritious.

A number of isolates of *Aspergillus ochraceus* from black pepper were grown in autoclaved moist corn and fed to ducklings, which are just about as sensitive to ochratoxin as they are to aflatoxin. Six of ten isolates from one sample of pepper when so fed resulted in death, in four days or less, of all the ducklings to which they were fed. It seems highly probable that some of the samples of pepper that were heavily invaded by *A. ochraceus* contained some ochratoxin, but, as with aflatoxin, as mentioned in chapter 3, we could not detect it, if it was present, because the solvents used to extract it extracted a good deal of other material from the pepper also, and we could not get rid of this "garbage."

Another fungus, *Penicillium viridicatum*, can also produce ochratoxin, and this fungus is relatively common in corn stored on the cob in cribs, at moisture contents above 20 percent and at low temperatures. This fungus will be taken up in more detail when the toxic species of *Penicillium* are discussed, but it seems probable that this fungus is a more common producer of ochratoxin than is *Aspergillus ochraceus*. When grown in pure culture in the laboratory it produces large quantities of ochratoxin, but only a few samples of ochratoxin-containing materials have been reported from the field — two of these were samples of corn of sample grade, which had a high percentage of damaged and decayed kernels and were obviously moldy and musty. It is highly unlikely that corn in that condition would be used for food anywhere. So far as I am aware, no single case of illness or death outside the laboratory has been attributed to ochratoxin poisoning, and it seems highly unlikely that ochratoxin poses any real and present danger to the health of people or to that of domestic animals. This may not be true of all regions at all times, but at least we do not know of any cases where its potential hazard has become actual hazard.

Other Toxin-Producing Species of *Aspergillus*

Several other species of *Aspergillus* are known or suspected to produce toxins, and these deserve a brief discussion.

The Species *Aspergillus Versicolor*

The fungus. Aspergillus versicolor is one of the storage fungi. It requires a moisture content in equilibrium with a relative humidity of about 80 percent to grow, which in the starchy cereal seeds is a moisture content of about 15.5 to 16 percent, wet weight basis. Evidently not much is known about the other ecological factors that influence its growth, but for one reason or another we seldom find it predominating in grains undergoing deterioration in storage; however, by the time the grain is obviously moldy, *A. versicolor* may be developing in it extensively, along with other group species of *Aspergillus* plus, usually, other filamentous fungi and yeasts. Some of the samples of pure black pepper mentioned in chapter 3 as being decayed by fungi were very heavily invaded by *A. versicolor*, but never by this fungus exclusively.

A few years ago an interesting case was described of heavy mold development on germinating barley in a malthouse in Scotland; the fungus grew so lushly on the germinating barley, and produced such a heavy crop of spores, that at times the workers who turned the malt with shovels could not see one another across the malthouse floor through the spore-filled air. The fungus turned out to be *Aspergillus versicolor*. That particular malthouse had been periodically plagued with this same problem. The owners and managers assumed that the offending fungus was being brought into the malthouse on the barley — that is, the fault could not be theirs, but must be someone else's. They got a mycologist working on the case to determine which lots of barley might be contaminated with *A. versicolor* so that they could avoid those. They evidently did not realize, or were unwilling to recognize, that the source of the trouble was not being brought into the plant from the outside but was simply lack of cleanliness within the malthouse itself. Everything in and around that malthouse must have been thoroughly and heavily contaminated by spores of *A. versicolor* (and probably by a lot of other fungi as well) so that whenever conditions were especially favorable it grew and prospered and sporulated mightily. It seems probable that malt as thoroughly invaded by this fungus as this malt was would have a decidedly moldy flavor; however, if it was used to make Scotch whisky, the mustiness probably was concealed by the peatsmoke flavor. The spores of the fungus that filled the air of the malthouse so that it could not be seen through must also have filled the lungs of the men who worked there, which probably would not add to their health or longevity.

The Toxin. When growing under the right conditions *Aspergillus versicolor* can produce sterigmatocystin, a toxic compound that was given that name because the fungus once was known as *Sterigmatocystis*. Sterigmatocystin has been incriminated as the cause of disease in calves that consumed feed heavily invaded by *A. versicolor*. In a subsequent experiment pure cultures of the fungus were grown on autoclaved feedstuff and this was admininstered to calves; they developed symptoms of the same disease, which is strong circumstantial evidence that the disease in the original lot of calves might have been the result of ingestion of sterigmatocystin. More clinching evidence would have been detection of sterigmatocystin in a sufficient amount in the suspect feed to have been responsible for the symptoms, but at that time this was not possible. Sterigmatocystin has been detected in one lot of moldy coffee beans in South Africa, but there is no evidence to suggest that even if coffee had been brewed from these moldy beans, the toxin would have appeared in the liquid, and it seems highly unlikely that anyone is going to go around munching on moldy coffee beans. The probability that sterigmatocystin will ever constitute a serious health hazard for man or beast seems remote.

The Species *Aspergillus Clavatus*

The fungus. Aspergillus clavatus requires a high moisture content to grow — free water or close to it — and so it is not a storage fungus in the sense in which this term is ordinarily used. It is common in soil and in manure and in decaying plant material — the favored substrates of so many fungi, including other species of *Aspergillus*. It sometimes is isolated from hay, seeds, or other feeds or feed ingredients, but usually only when these are in the final stages of decay.

The toxin. Aspergillus clavatus produces one or more toxic compounds that have been referred to as antibiotics when they were used to control or inhibit the growth of bacteria, and as mycotoxins when they have caused poisoning of animals. In at least one case the toxin produced by this fungus was proven beyond all reasonable doubt to have been responsible for the death of several people in a family in Thailand. After a portion of cooked rice had been eaten by the family, the rest of the dish was held without refrigeration and eaten on subsequent days. *A. clavatus* and the toxin produced by it were isolated from the rice; the fungus was also grown in pure culture and found to produce the same toxin as was found in the rice. The practice of preparing enough cooked, fungus-susceptible food for several days must make for high mycotoxin risk. The cooking destroys the natural mixed fungus flora in the material, and if it chances to become heavily contaminated with a toxin-producing fungus, that fungus can develop as a practically pure culture, which seems to be one of the requirements for toxin production by most fungi that produce toxins.

The Species *Aspergillus Fumigatus*

The fungus. Aspergillus fumigatus has long been known as a parasite of animals, and it will be met with again in chapter 6. Like so many of its relatives it is common in soil and in plant materials undergoing microbiological spoilage. If it is not truly thermophilic, depending partly on the definition of what constitutes a thermophile, it is at least thermotolerant; when growing vigorously in piles of horse manure or in mushroom compost, it can raise the temperature up to 50° C (122° F) in a couple of days, and hold it there for some time. During the stage of development of "hot spots" in stored grains when they are at that temperature, *A. fumigatus* may be a predominant organism. Steaming of soil or of compost will kill most fungi, but may only stimulate the spores of *A. fumigatus* to germinate later; the spores also can survive exposure to alcohol and to some other disinfectants that would kill most other fungi; *A. fumigatus* is a tough customer.

Toxins. Aspergillus fumigatus produces several toxic compounds that have been variously described as antibiotics or as mycotoxins, depending on the major interest of whoever was doing the describing; one of these compounds, named fumagillin, has been a sufficiently effective amoebicide so that it has been manufactured and sold under several trade names, but it has to be administered with care because only a little bit more than is needed to get ride of the amoebas will get rid of the patient too. Outside of the laboratory *A. fumigatus* predominates for the most part only in materials in the last stages of decay, by which time they look, smell, and taste so horrid that they hardly would be eaten in any quantity by animals or man.

Other Species

Several other species of *Aspergillus* have at times been implicated in outbreaks of disease associated with the consumption of moldy feed, but there has been no good evidence to establish a direct and unequivocal cause-and-effect relationship between consumption of feed invaded by a given fungus or group of fungi and the development of a given disease syndrome in the animals.

The Genus *Penicillium*

Taxonomy and Ecology of *Penicillium*

According to *A Manual of Penicillia* (48), published in 1949, there were at that time about 140 species in the genus. By 1968, 113 additional species had been added (46) by various mycologists and others around the world. Each of these species, of course, must be made up of many races, forms, and varieties that grade more or less insensibly into one another in the characteristics used to identify them. Theoretically, before one foists a new species onto a world already sorely overburdened with poorly defined species, it is desirable to become familiar with all the forms of all the closely related species that might possibly be confused with this new one, and to determine the range of genetic and environmental variation within those forms as well as in the forms and varieties of the new species. Depending on how thoroughly this is done, it can be a fairly formidable undertaking. For an ordinary mortal in an ordinary working life of, say, 30 to 40 years, to become thoroughly familiar with the total range of genetic and environmental variability in this complex jungle and jumble of species, so that he would KNOW that the species he is describing as new really IS new, is impossible. Either some of the describers of new species of *Penicillium* are not ordinary mortals (which they probably would be the first to admit) or some of these species are of questionable validity (which the describers of them would be the last to admit). Some mycologists struggling with the identification of this or that species of *Penicillium* work for days or weeks to identify it, and then are not too sure; others can identify just about any and all species of *Penicillium* readily, rapidly, and with great assurance. Similarly, to some dedicated describers of new species, just about anything they come across that is new to them, and that is different in any character of any kind, even the most minute or abstruse, is a legitimate new species, and is so described.

Some of the species of *Penicillium* must be counted among the most successful organisms on earth. As stated in *The Molds and Man* (42),

"Penicillium . . . can subsist on the remains of thousands of different kinds of plants, on cloth, leather, paper, wood, tree bark, cork, animal dung, animal and insect carcasses, ink, syrup, seeds of all kinds, manufactured cereal products and the boxes in which they are packed, including the wax and ink on the outside, on stored fruits and vegetables, soil, glue, paint, liquid drugs, hair and wool of all kinds, on the wax in our ears, and on literally thousands of other common products." Some cheeses such as Roquefort and Camembert and their various congeners are ripened with the aid of *Penicillium*; now selected strains are used for this, but until relatively recently this ripening was achieved by aging the cheeses in an environment where a given species of *Penicillium* was likely to predominate, and then there must have been a mixture of strains and probably of species present. None of the species of *Penicillium* can thrive at so low a moisture content as can some species of *Aspergillus* — the lower limit of moisture for the most xerophytic species of *Penicillium* (that is, those best adapted to limited water supply) is that in equilibrium with a relative humidity of about 85 percent — but some of them can grow at temperatures down to freezing or just below. They do not grow rapidly at this low temperature but, because they have few competitors there, they may grow as almost pure cultures, and, as noted above, if a fungus is capable of producing toxins, it is much more likely to produce them when growing in pure culture than when growing as one component of a mixed flora of microorganisms.

Toxicity of *Penicillium*

The possibility that rice invaded by some kinds of fungi, and especially by some species of *Penicillium*, might be toxic was investigated in Japan before 1900. At that time beriberi was a common disease in the Far East, and it still is encountered there as well as in other countries, including the United States. Beriberi supposedly is brought on by a lack of sufficient thiamin in the diet, and this in turn results from a dependence on polished rice. As late as 1969, however, Uraguchi (40), an investigator of competence, maintained that the cause of beriberi most likely was a toxin produced by *Penicillium toxicarium* growing in the rice. According to his account, in 1910 a program of rice inspection was adopted in Japan (it was adopted because someone recognized that moldy rice might not be wholesome) and from that time on moldy rice was excluded from the markets. The incidence of beriberi dropped dramatically. This was a year before the discovery of vitamins, and ten years before vitamins were used therapeutically in medicine. Uraguchi isolated the *Penicillium* toxin, administered it to rats, and produced symptoms similar to those observed in humans affected with beriberi. He also inoculated rice with the toxin-producing *Penicillium*, fed this to rats, and produced the same symptoms. This suggests that consumption of foods made toxic by growth in them of *P. toxicarium* may be *one* cause of beriberi — it does not exclude other possible causes.

Carlton *et al.* (41a) at Purdue University have done a good deal of work with *Penicillium viridicatum*; they were especially interested in this species because they had found it to be common in corn stored on the cob in cribs and in shelled corn stored in commercial bins, especially in corn of moderately high moisture stored at low temperatures, which is the condition of much corn throughout the winter in the corn belt of the United States. They grew two isolates of *P. viridicatum* on autoclaved moist corn for two weeks at 24° C (75° F), then dried and ground the corn and incorporated it into rations as various percentages of the total and fed these to albino mice. Both isolates resulted in a lack of weight gain, in deaths of many mice, and in the development of lesions in the internal organs, including the liver and kidney.

Krogh and Hasselager (45) in Denmark, say, "Kidney diseases in domestic animals, caused by feeding mouldy feed, have been known for a long time." They continue, "Mould nephrosis is widely distributed among pigs in Denmark, and occurs with specially high frequency after wet harvesting periods." They found two samples of moldy barley that were associated with kidney toxicity in pigs, fed these again to pigs, and produced the disease. They isolated various fungi from the suspect barley, including *Penicillium viridicatum*, grew these in heat-sterilized moist barley, and fed the barley to rats. Four of the ten isolates of fungi so tested caused injury to the rats, and the kidney lesions produced by oral administration of the *P. viridicatum*–infected corn to rats closely resembled those found in pigs that had eaten the toxic barley. This is strong, if still circumstantial, evidence that *P. viridicatum* not only was involved in but was responsible for the symptoms in the pigs. A further step would be to isolate and identify the toxin or toxins produced by *P. viridicatum*, and then detect this toxin or these toxins in the suspect barley. This will come, but it takes time and talent and money. However, if I were a pig farmer, I do not think that I would feed to my pigs any corn or barley or other grain heavily invaded by *P. viridicatum*; if I could not identify *P. viridicatum*, as most pig farmers probably cannot (a lot of mycologists probably cannot, either), I would try to avoid feeding *any* heavily molded grain to my pigs.

The case of moldy corn toxicity in Denmark was described in some detail to show what must be done to establish with even a reasonable degree of probability that a given fungus in a given lot of feed is responsible for a given disease syndrome in the animals that consumed the feed. Samples of silage, hay, corn, or other feed ingredients very heavily invaded by a variety of fungi, including *Penicillium*, occasionally are sent to our mycotoxicology laboratory at the University of Minnesota with a notation to the effect that the animals (usually expensive dairy or beef cattle) that have consumed this feed are not doing well, or that several have died, and asking us to "analyze" the feed to determine if it is toxic. The only way to determine whether such feed is injurious to the animals is to do as the investigators in Denmark did — feed it to animals of the same kind as it was fed to on the farm. Then, to determine whether any of the several kinds of fungi present might be responsible for the toxicity, it would be necessary to isolate these fungi, grow them in pure culture, combine this material in rations, and give these to test animals, including the same kind of animals as developed symptoms of illness on the farm. Each of the rations containing a different fungus should be fed to several animals, and of course there should be several control animals too. And preferably the tests should be replicated in time. In other words, to determine if a given batch of moldy feed is or is not toxic, and to find out why it is toxic (if it is) requires a moderately extensive and expensive research project.

The Genus *Fusarium*

Taxonomy and Ecology of *Fusarium*

Species of *Fusarium* are widespread in nature, growing as saprophytes in decaying vegetation and as parasites in the roots, stems, leaves, and fruits and seeds of wild and cultivated plants, in which they cause diseases such as wilts, blights, and rots. Many of these diseases are of great economic importance in some of our major crop plants over much of the world where these plants are grown, and for this reason a good deal of work has been devoted to *Fusarium* by plant pathologists and mycologists, including work on the taxonomy, classification, and nomenclature of the species. Several of the species of *Fusarium* that grow commonly in plants in the field or on ear corn stored in cribs produce some very potent toxins, and so *Fusarium* is of interest in mycotoxicology too.

The genus *Fusarium* was first described in 1808, by a German, some years before fungi were generally recognized as living things and about 50 years before they were shown to cause plant disease. Students of the genus have been describing new species of *Fusarium* ever since, and doubtless will continue to do so, probably under the impression or delusion that in so doing they are Contributing to Knowledge and are Pushing Back the Barriers. Sometimes they contribute to knowledge, sometimes they just contribute to confusion.

By the early 1930s more than 1000 species of *Fusarium* had been described, most of them identifiable, if at all, only by those who had described them. These were condensed by a couple of hard-working and methodical Germans into 65 species and 78 named varieties, many of them separated from one another on the basis of very minute differences. Shortly thereafter a new system of classification within the genus was published in an article by Snyder and Hansen of the Department of Plant Pathology of the University of California at Berkeley (52), and although few readers of this book are likely ever to be interested in the intricacies of *Fusarium* taxonomy, their approach is worth getting acquainted with because it could well be applied to the classification of so many kinds of living things. They collected cultures of different species of *Fusarium* from laboratories pretty much throughout the world, and from each of these cultures they made hundreds of single spore isolations and studied the resulting cultures in detail. If this was drudgery, it was well-conceived and meticulously carried out drudgery, and there was no other way in which the task could have been accomplished. Essentially, they used the experimental approach to determine the range in genetic and environmental variation within the species and varieties of *Fusarium*, and added to this more than a modicum of sound judgment and common sense. This is an approach that many other taxonomists might follow, but which few do — in part because it simply involves too much work, in part perhaps because, although they are biologists, they still do not realize the truth of what Charles Darwin said over a hundred years ago, that all species are varying in all characteristics at all times and everywhere. Snyder and Hansen reduced the number of species of *Fusarium* to nine, all of them identifiable by morphological characteristics of spores produced in culture on standard agar media. With their brief and clear descriptions (52, 53, 54), and the pictorial guide by Toussoun and Nelson (56), a nonspecialist now has at least a fighting chance of identifying a species of *Fusarium* in which he is interested. If he wants to be *really* sure, he had better send off a culture to Snyder or to Nelson for an authoritative diagnosis — both generously offer this valuable service without charge.

The *Fusarium* taxonomy war, however, is by no means over, and skirmishers continue to advance, fire, and fall back. At least a half dozen other systems have been devised by or revealed to fusariologists in different countries, most or all of these systems so involved and complex that only their perpetrators can understand and use them. A devious mind demands a devious system.

Toxins

Several species of *Fusarium*, as mentioned above, produce toxins, and two of these will be described in some detail.

Fusarium tricinctum. In the early 1940s there were a number of outbreaks of alimentary toxic aleukia (frequently abbreviated to ATA) in man in the USSR, especially in the Orenburg District. Joffe (44) says, "In 1944, the peak year, the population in the Orenburg and other districts of Russia suffered enormous casualties . . . more than 10 per cent of the population was affected and many fatalities occurred in nine of the fifty counties of the district." *Alimentary toxic* refers to the fact that the toxin is taken in by way of food, and *aleukia* refers to the greatly reduced number of leucocytes or white blood cells in the affected persons. Millet formed a great part of the diet of the people in the region of the outbreak, and at times much of the millet was left standing in the fields over winter because bad weather in the fall had prevented its harvest at the proper time. During late winter and early spring the millet was invaded by a variety of fungi, including *Fusarium tricinctum*. The people gathered and ate this fungus-invaded millet (not that they preferred it — they had nothing much else to eat) and many of them came down with what was diagnosed as ATA. Thousands were affected, and many died. At that time the cause of the disease was not known, but some expert field and laboratory work by Joffe finally pinned the blame on *F. tricinctum*. (He called it *F. sporotrichioides*, which Snyder and Hansen include in the *F. tricinctum* group.) He did not isolate and purify the toxin, but he did establish without question that a toxin produced by this fungus growing in millet in the field was responsible for the outbreak of disease in people who consumed the millet. This was almost 20 years before mycotoxins became big news, and Joffe's excellent work remained relatively unknown until about 1965, when he presented a summary of it at a symposium on mycotoxins at MIT.

Fusarium tricinctum appears to be a fairly important fungus in mycotoxicology in the United States too. Shortly after our mycotoxin work got under way at the University of Minnesota in 1963, we isolated *F. tricinctum* from a number of samples of crib-stored corn that were suspected to have been involved in illness of dairy cattle; in some cases the fungus grew from nearly 100 percent of visibly decayed kernels plated on agar. Those working on mycotoxins at the University of Wisconsin, where more research has been done on the toxicology of this fungus than has been done anywhere else, reported that *F. tricinctum* was one of the most common fungi isolated from corn suspected of being toxic to farm animals, especially cattle.

For maximum toxin production the fungus should be grown on autoclaved moist grain for a couple of weeks at a temperature of 25°–30° C (77°–86° F) so that it becomes well established, followed by a couple of weeks at 10°–14° C (50°–57° F) at which temperature most of the toxin is produced. There is a wide leeway in the incubation schedule at the different temperatures, the general requirement being that, once the fungus is well established, a period of fairly low temperature, or a succession of periods of fairly low temperature, will result in production of the toxin. Weather such as this occurs commonly in late winter and early spring in many regions.

Smalley *et al.* (50) tried various methods of assaying the toxicity of isolates of *Fusarium tricinctum*. They grew the fungus in natural or artificial media, extracted it with ethyl acetate as a solvent, condensed this, and applied it to the shaved skin of white rats. Of 29 isolates of *F. tricinctum* so tested, 22 caused a severe reaction or death, which is a very high proportion of toxic isolates. They similarly tested 41 isolates of the fungus from field corn, sweet corn, carnations, oats, sorghum, wheat, fescue (a grass), hay, turk grass, and cranberries, and 31 of these caused a severe reaction in the rats or their death. This ''skin test,'' by the way, was developed by mycotoxicologists in the USSR in the 1940s; it is a very good test for some kinds of mycotoxins but not for others, since some of the mycotoxins are lethal if taken by mouth but are not at all or only slightly toxic when applied to the skin — and vice versa. So if one is investigating possible toxins in a given lot of feed, regardless of whatever other tests are made, there must be some tests in which the material actually is administered as feed, since that is about the only way the stuff can get into animals on the farm.

Christensen *et al.* (43) grew *Fusarium tricinctum*, isolated originally from moldy corn suspected of being toxic, in autoclaved moist corn, then combined this in various proportions in otherwise balanced rations and fed these to young turkeys. The ration containing only *1 percent* of the fungus-invaded corn resulted in lack of weight gain; it also caused a peculiar and severe inflammation of the edges of the beaks of the birds. The ration containing 2 percent of corn invaded by *F. tricinctum* caused death of some of the birds and severe reduction in weight of the surviving birds, and also induced the same inflammation of the edges of the birds' bills. The rations containing 5, 10, and 20 percent of corn invaded by the fungus were lethal within a few days to the birds that consumed them.

The toxin produced by *Fusarium tricinctum* has been isolated, purified, and characterized; it has been given the common name of T-2. Fed orally to rats, it has an LD_{50} of 3.8 mg/kg (= 3.8 ppm), which is lower than that of aflatoxin, but still toxic enough. Once this toxin is extracted and purified in the laboratory, it is handled with respect.

According to evidence from tests in which small amounts of the toxin were fed for some time to rats, it is not carcinogenic or cancer causing.

Fusarium roseum. Probably even most city folks know that corn or maize is a major ingredient in rations for swine — as indeed it is in the rations for many other kinds of domestic animals, in many countries; and in many lands it constitutes a chief food for man also. Like many other grains and seeds, corn is subject to invasion by many kinds of fungi, both before and after harvest, and some of these may affect its value for food or feed. Two of the fungus-related problems in corn are the so-called refusal factor and the estrogenic syndrome; both of these have long been known, but only recently have we begun to learn enough about their cause to know what is involved. Both of these are important enough to justify the discussion given to them here.

If the weather is rainy as the ears of corn are maturing in late summer and early fall, *Fusarium roseum* may infect them, the infection beginning at the tip of the ear and progressing downward toward the base. (See figure 4-2.) The infection may be restricted to only a very few kernels at the tip of the ear, or it may involve all the kernels on the upper one-fourth

Figure 4-2. Cobs of corn infected by *Fusarium roseum*

or one-third of the ear. Whatever portion of the ear is involved, all the kernels in that portion are heavily invaded and partly decayed by the fungus. *F. roseum* sometimes produces a "perfect" or sexual or ascospore stage, and then it is known as *Gibberella zeae*, and the infected corn sometimes is called "Gibberella corn" or "Gib corn." This fungus-infected corn is unattractive to pigs, and they refuse to eat it — whence the term "refusal factor corn." Whatever the other ingredients, if the ration contains more than 5 percent of kernels with this refusal factor, pigs will not eat it; they will starve rather than consume it. Under some combinations of circumstances that prevail here and there in the field, infected corn contains an emetic compound produced by the fungus, and if this corn is consumed by swine they suffer prolonged vomiting, after which they sensibly refuse to eat more of the corn.

This refusal factor corn can be a rather serious problem. In the fall of 1972, infection of corn in the field by *Fusarium roseum*, with consequent development of refusal factor corn, was common in the northern portion of the United States corn belt from Nebraska to Ohio. In northern Indiana and southern Michigan refusal factor corn was prevalent on a high percentage of the farms where corn was grown, and in that region corn is a chief crop on most farms. Usually, of course, the fact that the pigs would not eat the corn, or the mixed feed containing the corn, was found out only when the feed was put before them. For a farmer who is raising 800–1000 pigs and has several tons of feed mixed up and delivered to his bins on Friday, and who then finds that the pigs will not eat it, the problem is a serious one indeed.

So far the only solution, and it is at best a poor and partial one, is to dilute the refusal corn with sound corn so that the pigs will eat it. First, of course, you have to find some sound corn nearby. Even then you cannot be sure that, just because the pigs will accept it, it will be wholesome for them and will enable them to gain weight normally.

Various treatments have been tried to make such corn acceptable — which may not mean making it nourishing — such as adding molasses to it to conceal whatever odor or flavor makes it unacceptable to the pigs, heating it, either dry or moist, in the hope of destroying or inactivating whatever it is that makes the pigs refuse it, and composting it to let it undergo microbiological heating. None of these treatments have made the corn acceptable to pigs; most of them are impractical anyway. The United States sells hundreds of millions of bushels of corn abroad each year — in the 1972–1973 season, probably about a billion bushels. Foreign buyers are very much aware of the refusal factor and of other mycotoxins in corn, and it behooves the seller, if he wants to prosper, to furnish corn free of these hazards. As this is written, the only way refusal factor corn can be recognized or identified is to feed it — or, more accurately, offer it — to pigs. If they refuse it, it contains the refusal factor. Or at least it contains *a* refusal factor. Pigs are moderately intelligent animals — they can be traumatized by various things such as being moved to a new pen, strange surroundings, or being offered a different ration, and they can go "off feed" for many different reasons. If they refuse a ration containing suspect corn, is it because the corn contains the refusal factor, or is it because of some other and more obscure reason? In most cases, if the corn contains as much as 1 percent of kernels damaged sufficiently by *Fusarium roseum* to contain the refusal factor, this probably could be recognized by visual inspection by someone familiar with the type of damage caused by this fungus. Plating out the kernels on agar to detect the presence of the fungus is of little or no value, as is true of other mycotoxins as well. This point seems difficult to get across to veterinarians and feed manufacturers and farmers, probably in part because in the diagnosis of most infectious diseases, the final proof that the disease is caused by this or that bacterium or virus is the isolation of the causal organism. But these mycotoxicoses are *not* infectious; they are caused by a toxin or toxins excreted by the fungus into whatever the fungus is growing on. The fungus may die, but the toxin remains, so the absence of the fungus tells us nothing about the presence or absence of the toxin. Conversely, since any toxin-producing fungus requires rather special conditions to produce a biologically significant amount of toxin, the presence of the fungus does not mean that toxin is present. As noted in chapter 3, even when we heavily inoculated a sample of grain with a potentially toxic fungus, and let it grow along with the other fungi naturally present, no toxin was produced. The *F. roseum* that produces the refusal factor dies out in shelled corn within a few months after harvest, and then its presence cannot be detected by any means whatever, although the refusal factor in that corn remains as potent as ever. The only ways that any mycotoxins can be detected in grains or feeds or any other substrate are (1) to isolate the toxin, purify it, and identify it by spectrographic or other analysis; (2) feed the material to test animals and see if they develop the symptoms characteristic of poisoning by that particular toxin. The compound or compounds responsible for pigs refusing to eat corn invaded by *F. roseum* have not yet been isolated and characterized; the only way we can detect refusal corn, or detect refusal feed made from such corn, is to offer it to swine. If they refuse it, it is refusal corn or refusal feed. Determining whether the fungus is or is not present is a waste of time, and it tells us nothing. No one has yet been able to consistently produce the refusal factor in the laboratory; until that is done, it is necessary to depend on refusal corn produced by nature in the field, and since that happens only sporadically, those interested in the problem work on it only sporadically.

The estrogenic syndrome in technical terms involves, in female swine, swollen and edematous vulvas (vulvovaginitis); vaginal or rectal prolapse; edematous and tortuous uteri; atrophic, shrunken, and nonfunctioning ovaries; abortion; and resorption of fetuses. (See figure 4-3.) In the male it involves feminization — atrophy of the testes and enlargement of the mammary glands. Just a recital of the symptoms takes one aback somewhat. This estrogenic syndrome has long been an economically important disease in swine just about everywhere that swine are grown, not only in the United States but throughout Europe and the USSR. It is more prevalent in some places and in some seasons than in others, of course, but it is encountered in some herds here or there every year.

From the late 1920s on there were occasional suggestions that estrogenism in swine resulted from eating moldy feed, but none of those who mentioned this possibility had any experimental evidence to establish a cause-and-effect relationship between a given lot of moldy feed, or a given mold or fungus, and the symptoms in the swine.

The proof came in 1965, when the fungus *Fusarium roseum* was isolated from corn suspected to have been involved in an outbreak of estrogenism in a herd of swine; the fungus was grown in autoclaved moist corn, this was fed to pigs, and they developed the estrogenic syndrome. Since this was a mild breakthrough of sorts, and there may have been a bit

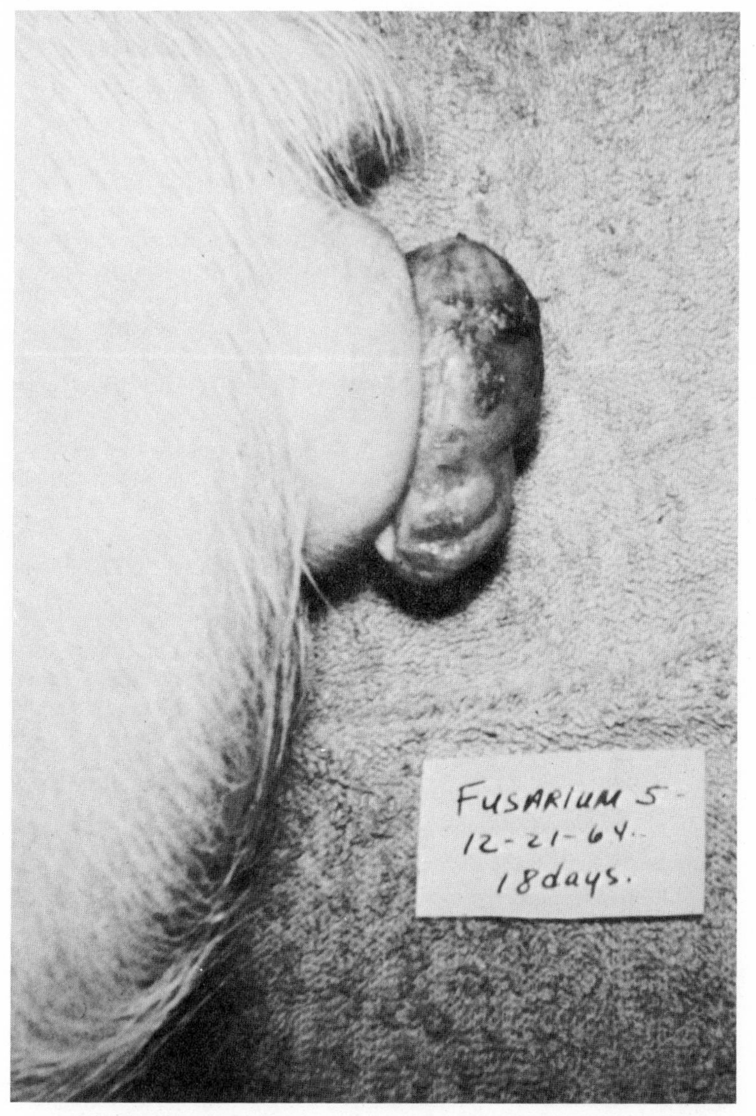

Figure 4-3. Vaginal prolapse resulting from consumption of corn in which *Fusarium roseum* had grown and had produced the estrogenic compound F-2

more involved than meets the outsider's eye, perhaps a short quotation from the paper reporting this experiment (47) is not out of order.

> Our evidence indicates that a common and perhaps major cause of this disease is a metabolite produced by the fungus *Fusarium*. Corn stored on the cob in cribs commonly is invaded by *Fusarium*, as well as by other fungi. Of the 75 isolates of *Fusarium* from crib-stored corn that we have tested, 35 when fed to rats have resulted in pronounced increase in weight of uteri in the rats in four to seven days. Some of the isolates, when grown in autoclaved moist corn and fed to rats, have resulted in death within four to five days, with increases of up to 10 times in weight of uteri. . . .
>
> Abortions in swine usually are attributed to leptospirosis, for the diagnosis of which a serological test is available. We commonly encounter a high incidence of abortion in herds that are negative for leptospirosis or brucellosis or both. These abortions are diagnosed as idiopathic; that is, of unknown cause. In one of our tests, four purebred Yorkshire gilts, all immunized against hog cholera, erysipelas and leptospirosis, and which at the start and finish of the test were negative for brucellosis and leptospirosis, were fed as follows: the controls received normal sow ration; the others received feed containing, respectively, 25, 50, and 100 per cent corn invaded by the fungus. Four days after the test began the gilt receiving a ration containing 50 per cent of corn invaded by *Fusarium* developed symptoms of the estrogenic syndrome, and aborted after 21 days. The control gilt weaned a litter of 10 pigs, and the three gilts fed different amounts of corn invaded by *Fusarium* weaned a total of 11, or an average of 3.7 per gilt.

The main estrogenically active compound has been isolated and identified, and methods have been developed to detect relatively minute amounts of it in corn or in pelleted or otherwise processed feed. The procedures for this still are too involved to have become a standard part of evaluation of batches of feed; a simple, one-step test to detect the compound, called F-2, would be invaluable to feed manufacturers and swine growers.

Note that in the account above it was stated some of the rats fed on corn invaded by *Fusarium roseum* died within four days. Some isolates of *F. roseum*, in other words, produce lethal as well as estrogenic compounds. We have some isolates of *F. roseum* from crib-stored corn that are as lethal, when consumed, as any fungus that we have tested. *F. roseum* is not just *F. roseum*. And although *F. roseum* is the principal producer of

F-2 among the fungi that we have tested, other species of *Fusarium* do produce lesser amounts of it, although seldom, in our experience, do they produce enough F-2 or related estrogenic compounds to be of much importance in the field. There is still another complication, in that estrogenic compounds produced by these fungi are not the only causes of the estrogenic syndrome in swine; a number of cases have been encountered in which estrogenism in large herds of swine was traced to contamination, by diethylstilbestrol, of pelleted or ground swine feed, presumably in the feed mills. This was for years added to feed manufactured for beef cattle, to hasten growth, and it seems probable that the mixing machines were not scrupulously cleaned after such a batch of feed was mixed and before a batch of feed was mixed for the swine. In one such case the man who owned the feed mill also owned the herd of swine — and was a practicing veterinarian to boot! The addition of diethylstilbestrol to feed for beef cattle was banned as of January 1, 1973, and so this should eliminate the man-made estrogen problem in swine; the estrogenic syndrome resulting from consumption of corn or other feedstuff invaded by *F. roseum* probably will be with us forever.

It was stated above that *Fusarium roseum* may be lethal as well as estrogenic, that it produces different kinds of toxins. This was very clearly established by one of our feeding tests at the University of Minnesota. An estrogen-producing isolate of *F. roseum* was obtained from a sample of corn that was associated with a case of estrogenism on a farm in southern Minnesota. This isolate was grown in autoclaved moist corn for a time, and this corn was added to a ration given to young swine, in an amount to furnish 500 to 600 ppm of F-2. This corn was included in their ration for the first 64 days, then they were put on a ration of commercial pig feed for another 64 days, the aim being to see if, during the second period, they would entirely recover from the effects of the estrogen that they consumed in the first 64 days. They did recover from the estrogenic effects but, at the end of the test, the male and female pigs on the control ration averaged 163 and 157 pounds in weight, respectively, and the male and female pigs on the ration containing the corn invaded by *Fusarium*, for the first 64 days, weighed an average of 102 and 86.3 pounds, respectively. That is, something in the feed dwarfed them during the first 64 days, and they never recovered from that; there was something other than the estrogenic F-2 in this corn. Feeding them the fungus-invaded grain produced quite different results from feeding them a ration containing pure F-2. Some mycotoxicologists have maintained that only those feeding tests are valid that are made with the purified mycotoxins added to the ration, that all feeding tests made with fungus-invaded grains are of no value whatever. They do not say how they came by that, to my view, totally erroneous, even witless, idea. On the farm, or in the zoo, or wherever mycotoxicoses are of economic importance, they occur as a result of the animals consuming feeds invaded by toxin-producing fungi, not feeds made toxic by the addition of purified mycotoxins. Both approaches can be used, of course, but if we are to understand what happens to the animals in the field or in the feeding pen, we had better work with some feeds invaded by fungi in the way they are in nature.

Other Mycotoxins

Stachybotrystoxin or Stachybotryotoxin

Stachybotrys is common in soil and on cellulosic materials such as straw, and when growing on straw under the right conditions it can produce several potent toxins. In the USSR and in Eastern Europe straw is commonly used as part of the feed for horses during the winter, and if the straw contains *Stachybotrys* toxins the animals that consume it may come down with either chronic or acute poisoning, depending upon how much of the toxins they ingest. The disease was recognized as a separate entity in the USSR in the early 1930s, but the cause was not established until some ten years later. In some years there it evidently is prevalent enough to be of real economic concern; it occurs during the winter when the horses are fed in the barns; when they are turned out to pasture in the spring and no longer consume the straw made toxic by the fungus the disease disappears. It is said that pigs sometimes develop a skin rash from lying on straw containing *Stachybotrys* toxin. So far as I am aware, no field cases of stachybotrystoxicosis have occurred outside of Eastern Europe and the USSR, but this could be due in part to lack of identification of the disease. However, it is not a usual practice in many countries to feed straw to domestic animals, and if the toxin is found mainly in straw, as it seems to be, this could explain the restricted distribution of the disease.

Facial Exzema in Sheep

Pithomyces chartarum is a common saprophyte on dead grasses in various parts of the world, and in cool and rainy weather it may produce heavy crops of spores on dead pasture grasses. The spores contain a toxin named sporodesmin (the fungus was first identified as *Sporodesmium*) and if a grazing sheep ingests several grams of spores (which is really a large quantity of spores), poisoning may result, the outward sign of this poisoning being facial exzema, although inwardly the chief damage is to the liver. The chemical process is interesting: One of the normal products of digestion of chlorophyll is phylloerythrin, a photosensitizing compound — it makes the skin sensitive to light. The livers of healthy sheep remove this compound from the blood, which is what the liver is supposed to do. The livers of sheep damaged by sporodesmin from the spores of *P. chartarum* do not destroy or excrete this compound, but pass it into the bloodstream; it accumulates in the capillaries in the outer skin and, where the skin is exposed to light, as on the face, it results in exzema. The disease has at times been prevalent in the North Island of New Zealand, and it occurs occasionally in portions of Australia. One means of control is to keep the sheep from pastures where the fungus is sporulating heavily. Spore traps have been developed that enable the investigator to determine the number of spores of *P. chartarum* per cubic foot of air in the field, and when the spores reach a concentration of about 10,000 per cubic foot of air in a given pasture the sheep are removed to feeding lots until the spore concentration drops below the danger point. A combination of irrigation, to make the grass grow more lushly, and of heavy grazing, to forage off the grass close to the ground, reduces the amount of dead leaves and stems and this reduces the material on which the fungus can grow, and so reduces the danger of poisoning.

The Slobber Syndrome

This inelegant but descriptive name refers to a disease of which the most obvious symptom is excessive salivation, although a variety of other symptoms may be involved. This results from consumption of red clover on which the fungus *Rhizoctonia leguminicola* has grown.

Detection of Mycotoxins

The reader may well ask, if these mycotoxins are as important as they are said to be, and as they seem to be, in the health of domestic animals and man, why more specific cause-and-effect relationships have not been worked out between a given fungus and a given disease or disease complex. And why, when animals in a given herd or flock are ill or dying, and all known causes of disease have been eliminated and the heavily molded feed is suspected — and given to mycologists to test — these scientists still cannot give a clear-cut answer whether the moldy feed is or is not implicated. The researchers ask themselves the same questions, as do those who support the research, and as do many in the food and feed industries who want the answers but who adamantly refuse to contribute financial aid to help pay for the work that might provide the answers.

As indicated above, the problem of mycotoxins and mycotoxicoses is not like the problem of a given bacterium or fungus or virus infecting a given animal and causing a specific disease. In that instance, if you can isolate the bacterium or fungus or virus, grow it in pure culture, inoculate it into test animals and produce the disease, then reisolate it again from the diseased animal, you have practically won your case.

Foods and feeds invaded by fungi that are potentially toxin producing, however, present a different problem — or at least many of the feeds do. They are almost always invaded by several to many kinds of fungi, plus an assortment of bacteria and occasionally mites and insects. This mixed and varied population is fluctuating, shifting, and changing from day to day, and sometimes from hour to hour. Suppose that some animals in a herd or flock become ill, and die. Several of the carcasses may be sent to the diagnostic laboratory of the nearest college of veterinary medicine, where they are examined for known disease-causing agents; if none are found, the feed may be suspected, and samples of the mixed feed and perhaps of the separate ingredients, are submitted for examination. Examined how, and for what? As was stated above, determining the number and kinds of fungi present tells us nothing about the presence or absence of toxins that might be produced by these fungi; Shank *et al.* (49) said, "Aflatoxin contamination did not correlate well with the presence of viable *Aspergillus flavus* in foods at the time of collection and, consequently, the presence of viable organisms was not a reliable index of aflatoxin contamination." Test the feed samples for the presence of known mycotoxins? This can be done, but only some half-dozen or so mycotoxins can be identified in this way, and these can't always be identified even if they are present. According to Smalley *et al.* (50), samples of baled hay that were suspected of being toxic ". . . contained fluorescent compounds with thin layer chromatograph (TLC) mobilities identical to aflatoxin B_1 which made analysis impossible." And even if a known fungus toxin can be isolated from a given lot of feed, this is no proof that this toxin was present in the feed consumed by the animals that became ill and died; the sample sent in for identification is not the feed that the animals consumed, but is only from the same lot, and no one knows to what conditions it may have been exposed in the meantime. The toxin may have developed in this sample *after* the animals died. Also, mycotoxins are not the only toxic substances that can get into feeds — grains treated with fungicides have been at times sold for feed, contrary to law and to common sense. A batch of turkey feed involved in the death of more than 6000 turkeys on a southern Minnesota farm in 1973 was found to contain four times the recommended amount of arsenic; someone at the feed mill was careless. Some of these highly toxic chemicals are relatively easy to detect, but others are not.

One approach to determining whether a given batch of feed is responsible for a given set of symptoms in animals is to segregate a few animals and feed them on the suspect ration for a time and see what happens. If this establishes the fact that the feed is indeed responsible, it can then be subjected to various analyses that will possibly establish the identity of the toxic component.

Numerous investigators have tried to accumulate information on possibly toxic fungi by growing the fungi in various substrates and then administering these moldy materials to animals of one kind or another. Sometimes the fungi have been grown in liquid media, and the medium or the fungus substance itself has been either added to a ration or administered by way of a stomach tube to laboratory animals. As an alternative approach, the fungi have been grown in pure culture on various grains, and these moldy grains have been fed to animals, or offered to them, as a portion of their ration or sometimes as their sole ration. In some of these tests this has resulted in the animals being given a ration far more heavily invaded by fungi than ever would occur in practice. Some of these heavily molded materials were so unattractive to the test animals that they refused to eat them, and died of malnutrition — bluntly, they starved to death, and yet the investigator did not mention this, or mentioned it only in passing. As an example of how misleading the results of some of these tests can be, the fungus *Cladosporium*, which is common in some grains, especially when they are exposed to humid weather in the field just before harvest, was grown in autoclaved moist corn for a time, then this corn was dried, ground, and fed to rats as their sole ration, and it killed them within two days, after they had consumed only a couple of grams of the moldy feed (a gram is about 1/28th of an ounce). The test was repeated several times, always with the same result. (These tests were made by our mycotoxin team at the University of Minnesota.) Later, the same isolate of the same fungus was grown in the same way, and again fed to rats, but this time as 50 percent of an otherwise well-balanced ration — and the rats gobbled it up and throve on it. In nearly all the tests of this kind, the fungi have been grown in pure culture. In the field they generally occur as mixtures, seldom as pure cultures. In chapter 3, tests were described in which it was shown that consumption of rations heavily invaded by a mixture of fungi, including a known potential toxin producer, resulted in no ill effects whatsoever; some feeds compounded in part of very heavily molded grains, in fact, resulted in better weight gain than did the control feeds compounded with food-grade corn.

The moral of this is that qualifications attach to all such tests with moldy feed ingredients, and especially to tests with pure cultures of fungi administered or offered in relatively massive doses. The results are affected by so many factors: the isolate of the fungus used; the substrate on which it was grown; the temperature and time of incubation; the kind of animal it is fed to; the amount consumed; the makeup of the ration. There probably are others, but those are the main ones, and investigators as well as the readers of investigators' papers should be aware of these qualifications. It is important to find out the conditions under which moldy or fungus-invaded foods and feeds are *not* harmful, as well as the conditions under which they are. Some fungi have been incriminated as possible hazards in foods and feeds on the basis of evidence from tests that have had only a tenuous relation to reality, to what happens in nature. This is not to question the fact that some fungi produce highly toxic compounds, and do so in the field and in the warehouse as well as in the laboratory — of course they do, and, when they do, they pose a health hazard to us and to our domestic animals, and it behooves us to learn all we can about them as soon as we can. It *is* intended to question the sometimes almost hysterical statements made by otherwise fairly reasonable people to the effect that probably *all* fungus-invaded or moldy foods and feeds constitute a serious health hazard; there is no evidence to support such a statement, and there is a good deal of both observational and experimental evidence to indicate that it just is not true. Eventually, say in a hundred years or so, many of the answers to many of the most pressing problems in mycotoxicology will be available. In the meantime, let us proceed with a blend of wariness and common sense.

5

Airborne Fungus Spores, Plant Disease, and Respiratory Allergy

That spores of some fungi could become airborne was shown by actual test nearly a hundred years ago, in 1882, when Marshall Ward, one of the early major prophets in plant pathology, caught spores of the coffee rust fungus, *Hemileia vastatrix*, as far as 25 feet away from the nearest coffee tree. Though this may not sound like an epochal discovery, in a way it was, because the spores of this fungus are peculiarly adapted to dissemination by splashing raindrops, not by wind. (As an aside, this fungus, a native of Africa, was recently detected in Brazil, something that has long been feared and that probably will turn out to be a Grade A calamity for Brazil and for other South American countries where coffee is grown. Were the spores blown across the south Atlantic from Africa to Brazil? It seems unlikely; someone, somehow, carried them.) Shortly after Ward's work, Millardet, in France, caught up to 32,000 spores of the grape downy mildew fungus, *Plasmopara viticola*, per square decimeter of surface in a grape orchard. From then on, a great many plant pathologists interested in the epidemiology of plant diseases became very conscious of airborne spores and devoted considerable time and talent to the detection of the fungus flora of the air. By the early 1900s the present Great Plains wheat belt had been established, with practically solid wheat from southern Texas to the Dakotas and Montana, and on hundreds of miles more into Canada. It was already suspected that at times urediospores of the black stem rust fungus — the ''repeating'' spores that spread from one wheat plant to another — were being carried northward in the spring from fields of rusted plants in the South to plants farther north, and plant pathologists began exposing microscope slides to the air outdoors and examining them for the characteristic spores of the rust fungus. In the 1930s some of the more perceptive allergists who had long been interested in airborne pollen as a cause of respiratory allergy, or hay fever, realized that airborne fungus spores might also be involved. Their interest, along with that of plant pathologists, resulted in the development of what became known as aerobiology — the study of living things, especially those of microscopic size, in the air.

Actually Marshall Ward was not the first to show that air might carry biological freight. Charles Darwin, of *Origin of Species* fame, when on the survey ship *Beagle* in 1832, collected dust from a sort of windsock at the masthead when the ship was somewhere out beyond the Azores on its way to Brazil. He examined the dust microscopically and found in it wheat he called "infusoria" — most likely what now would be called diatoms, which are microscopic single-celled algae with two shells or valves made up of silicon, giving the shells a high degree of durability; in places there are great deposits of these diatoms from ancient seabeds, known as diatomaceous earth or kieselguhr. (They are a main ingredient of polishing powders.) Darwin had to have the idea that such dust was worth collecting and examining, and then had to devise techniques to do this; at that time he was still a young man, with very little firsthand experience in observational or experimental biology. As anyone who has read the *Voyage of the Beagle* must realize, even at that early time Darwin would have to be rated as a Bright Young Man. It is perhaps unfortunate that few Bright Young Men of today, even Bright Young Biologists, read his works. Darwin supposed that these infusoria, which he collected in some quantity far from land, had been blown out from the Sahara Desert, a fairly long haul away, although the German savant who later identified some of them for him said that he recognized two species prevalent in South America but not known to occur in Africa, and South America was an even longer haul away. From the general west-to-east direction of the upper air currents in that region, South America seems to be the more probable source of this dust. Darwin says that other navigators had at times reported dust storms, or at least heavy deposits of dust, far out at sea.

Since about 1900 a great amount of work has been done in aerobiology in general, and in the dissemination of fungus spores by air, in particular — there must be some thousands of research papers devoted to different aspects of the subject, as well as a few books. What follows, then, is a summary of some portions of this extensive work that will serve to introduce the reader to the field in general; the references to this chapter at the end of the volume should enable those so inclined to delve into it further. For such readers, the book by Gregory (62) is highly recommended.

Numbers of Spores Produced by Fungi

A great majority of the more than 100,000 species of fungi produce at least one kind of spore that is adapted in one way or another to dissemination by wind; many species produce two or more kinds of spores that are disseminated primarily by the wind; and still other species produce spores that, although not especially adapted to wind dissemination, nevertheless are commonly carried by the wind. This is a chancy and haphazard method of distribution, of course, and for it to be as biologically successful as it is the production of relatively tremendous numbers of spores is required. In sheer numbers of reproductive units, no other group of living things can even come close to the fungi. A few examples will illustrate this.

Fomes applanatus is a wood-rotting fungus common in hardwood forests. The mycelium invades and decays the lower part of the trunk of many kinds of trees, and after it has grown in the tree for some years it produces fruit bodies on the lower part of the stem (see plate 7) near the ground, or the decayed tree may topple over and the fungus then produces fruit bodies on the fallen log, sometimes a dozen or more per log. The fruit bodies are perennial and continue to increase in size from year to year; those a foot across are not at all uncommon. The spores — basidiospores — are produced on basidia on the walls of pores or tubes on the underside of the shelflike fruit body. (See plate 8.) The tubes are small in diameter, just visible to the naked eye, and they average about 50 per centimeter, or 125 per inch. The walls between the tubes take up about half the space, and so each tube opening is about 1/250 of an inch in diameter. The tubes of the current year begin to grow in April or May, and continue to grow until October or November. During all this time basidiospores are produced constantly, mostly near the advancing lower end of the tubes. On a quiet, humid day the spores can be seen issuing from the fruit body in a thin, wavering, wraithlike cloud, thin, but dense enough so that it can be photographed. The cloud disappears a few feet away, at most, when the spores become so dispersed that the mass no longer is visible, and the tops of the fruit bodies almost always are covered with a powdery brown coating of spores that have been deposited there. It has been calculated that a fruit body of moderate size would, in a season of six months, produce 5460 billion spores, or 30 billion per day, or 1.25 billion an hour, or 21 million per minute, or 350,000 per second. Day in and day out, for six months. In old hardwood forests there may be several hundred fruit bodies of *F. applanatus* per square mile. It seems highly probable that whenever the environment is favorable to the establishment of *F. applanatus*, the fungus will be there. The same is true of just about all wood-rotting fungi.

The giant puffball, *Calvatia gigantea*, is well named, since it grows to a somewhat amazing size; specimens a foot in diameter are common, and occasionally one is found 18 inches in diameter and with a weight, when fresh, of more than 25 pounds. As the puffball matures the entire interior turns into a mass of spores and associated thin threads, the capillitium, among which the spores are entangled. The outer wall of the puffball is fragile and easily broken, and as it weathers away the spores are exposed to the air and many of them become airborne in one way or another. The spores are small — three to four microns, or 1/6000 to 1/8000 of an inch — in diameter. A fair-sized giant puffball has been estimated to contain 7,000,000,000,000 spores, or, expressed in another way, 7×10^{12}.

The pear-shaped puffball has fruit bodies only an inch or so in diameter, but it makes up for this relatively small size by producing them in clumps of a dozen or more, on decaying stumps and logs. When they are in season it is easy to collect a quart of them — they make excellent eating. As they mature a small pore, about ⅛ inch in diameter, forms in the center of the top, through which spores can escape. The wall of the puffball is relatively firm, but it will give a bit with a slight impact; a raindrop striking the fruit body bends the wall in slightly, and this bellows action is sufficient to expel a million spores. Many of these spores are washed out of the air before they have traveled far, but many of them are not. Just about anything that comes in contact with the puffball will similiarly cause it to puff out spores. That this method of dispersal is effective is indicated by the fact that pear-shaped puffballs of northern Europe and the USSR and those of the United States and Canada evidently are identical.

Mushrooms or gilled fungi also are prolific producers of spores. A single average-sized fruit body of *Agaricus campestris*, the wild brother of the common cultivated mushroom, can produce 16 billion spores, all within a period of about 24 hours. Many other gilled fungi and their nongilled relatives among the fleshy fungi are equally prodigal in spore production. The forest primeval, or as close as we can get to such a condition in these times, the wilderness areas, in rainy weather in the fall abound with hundreds of kinds of fleshy fungi plus thousands of kinds ot other fungi that grow on living, dying, and dead plants, on tree bark, on decaying wood, and on just about everything else living and dead in the forest. The air in such a forest may smell wonderfully fresh and clean and invigorating, but it is by no means pure; it bears a heavy if invisible load of spores of all these fungi. Fortunately, almost none of these ever have been implicated in respiratory allergy. As with pollen, so with fungus spores — a few kinds cause most of the trouble; the rest are innocuous.

Daldinia concentrica has no generally accepted common name. The fungus is a wood-inhabiting Ascomycete, and it produces hemispherical black fruit bodies, up to an inch or two in diameter, in late summer and fall. Usually several fruit bodies appear together on a decaying branch or log. The interior of the fruit body is composed of a series of concentric zones so designed as to get the greatest mass for the smallest weight, each of the zones of dense mycelium alternating with a zone of less dense mycelium. Just beneath the outer surface of the fruit body (technically it is called a stroma) the perithecia form, and within these develop the asci that contain ascospores. When the ascospores are mature they are forcibly discharged, but only at night, from about 10:00 P.M. to 5:00 A.M., with the peak discharge being at 11:00 P.M. According to Ingold and Cox (67) who studied this phenomenon, from 1500 to 5300 spores were discharged per square centimeter of surface of the stroma per 15 seconds at the peak time, with a total of 2,062,000,000 spores being discharged over the life of the stroma. Many of the spores land on the wood on which the fungus is growing, and make a black halo around the stroma, but many of them get into the air and are carried away. For more on this see the books by Ingold (65, 66).

The production and liberation of spores of many fungi are time-correlated, as in *Daldinia concentrica*, and in some of them this built-in biological clock cannot be altered by anything that we subject the fungus to in the laboratory; when that period of the day or night arrives at which the fungus is to produce and liberate spores, it produces and liberates spores. Spore production and liberation in other fungi are determined by temperature or by relative humidity, and some of these adaptations are very precise indeed.

In the Philippine Islands corn or maize often is infected with the fungus *Sclerospora philippinensis*, which causes downy mildew. This fungus, an obligate parasite, as are all downy mildew fungi, produces conidia or sporangia on branched stalks that grow out of the stomata, the minute openings in the epidermis of the leaf that permit gas exchange between the interior of the leaf and the surrounding atmosphere. There are several crops of sporangiophores, or spore-producing stalks, per night, each crop consisting of 30 or so sporangiophores per stoma, with 20 to 40 spores per sporangiophore. This may go on for two months, if the host plant lives that long. According to Weston (70) the number of spores produced by this fungus on a single plant per night ranges from a minimum of 758,033,400 to a maximum of 5,946,069,000, and in two months the fungus on a single plant could produce 500,000,000,000 spores. With 10,000 plants per acre and 640 acres per square mile, the fungus could produce 3,000,000,000,000,000 spores, or 3×10^{18} spores per square mile per season. This fungus produces its spores only at night, from about 11:00 P.M. to about 6:00 A.M. Weston tried in various ways to upset this schedule, including putting infected plants in a dark box with uniform temperature and humidity — but the fungus continued to produce its spore crops from 11:00 P.M. to 6:00 A.M.

Some other downy mildews also produce their spores only at night. One of these infects a wild grass with the generic name of *Setaria*, and both the grass and the mildew are common in Europe and the United States. This fungus had been known since about 1875, and had been worked over by mycologists in half a dozen countries, but no one who ever studied it was able to find normal spores, only dried up and misshapen ones; the general consensus was that it produced only aborted conidia or sporangia, in which case the question was how it could propagate itself so successfully. Weston solved this mystery very simply and easily — the fungus was sporulating at night; for 50 years the mycologists had been dormant when the fungus was working and working when the fungus was dormant. The moral of that, if one is needed, is that nocturnal things are best studied at night.

According to Stakman (69), in the 1925 epidemic of black stem rust of wheat in the Great Plains area of the United States, "it was calculated that in early June, 1925, there were 4,000 tons of urediospores (150 billion spores per pound) on four million acres of wheat in 16 counties of northern Oklahoma and south-central Kansas. Winds carried the spores to Minnesota and the Dakotas where they were deposited at the rate of 3.5 million an acre in an area of 40 thousand square miles. On an acre of fairly heavily rusted wheat there are about 50 thousand billion spores." That's a large quantity of spores. The figures above, by the way, were not just picked out of the blue, but rather the spores themselves were. Stakman and his co-workers in plant pathology at the University of Minnesota, where for many years most of the work on epidemiology of black stem rust of wheat was carried on, and where work still is carried on, had undertaken a survey of when and where rust spores were produced and where they were carried to by the air, since they recognized that such knowledge was necessary for an understanding of the epidemiology of the rust. As a part of this survey, glass microscope slides coated with a thin layer of vaseline were exposed by collaborators at many stations from the southern portion of the winter wheat range, in southern Texas, to the northern portion of the spring

Plate 1. Colonies of fungi growing from "pure" black pepper scattered on agar. The yellow colonies are *Aspergillus flavus*

Plate 2. Fungi — mainly *Aspergillus flavus* — growing from elbow macaroni plated on agar

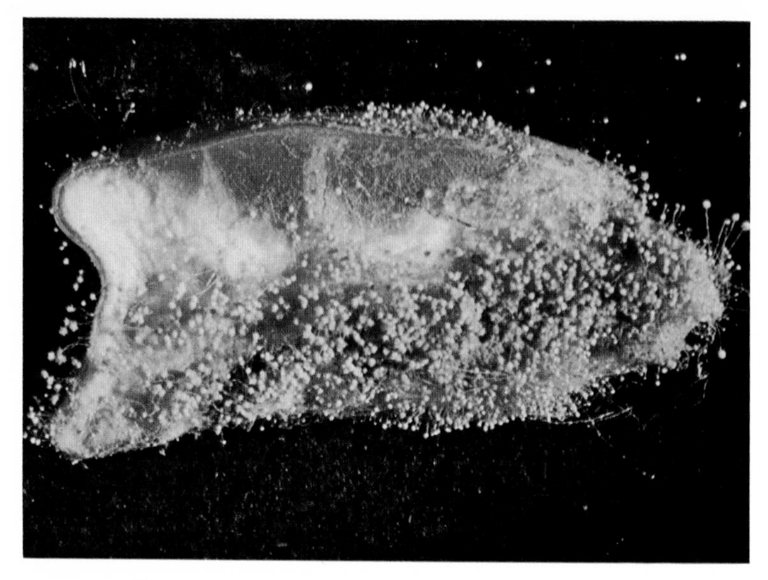

Plate 3. Several species of *Aspergillus* growing from a kernel of corn that developed damage in storage

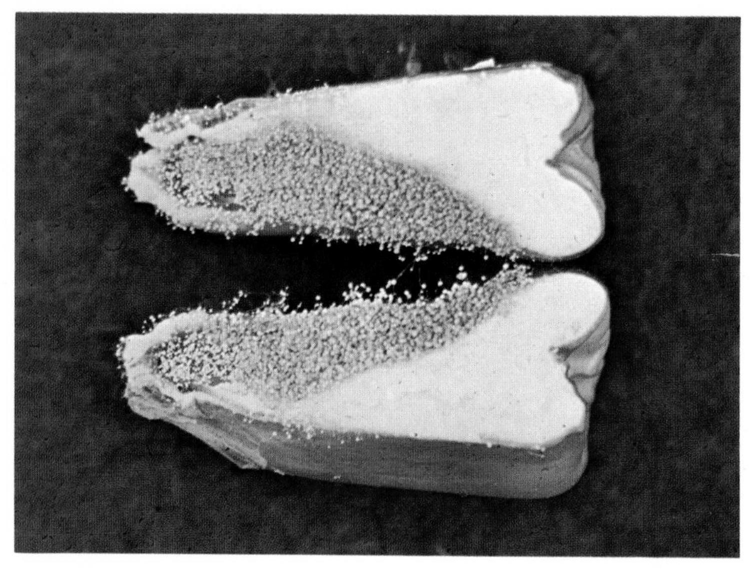

Plate 4. *Aspergillus flavus* growing from the germ or embryo of a split kernel of corn that had been inoculated with the fungus

Plate 5. A halo of spores of *Empusa muscae*, shot off after the fungus had invaded and killed the fly

Plate 6. *Coprinus atramentarius*, one of the mushrooms that is toxic to some people if consumed along with alcohol

Plate 7. *Fomes applanatus*. The brown color on the surface of the fruit bodies consists of spores released from pores on the underside

Plate 8. The yellow deposit on the surface of the log, and on some of the ferns, consists of spores of the fungus on the underside of the log

wheat belt, in North Dakota. The slides were exposed daily, usually on the front of a wind vane so that the slide was held into the prevailing wind; spores (and a lot of other material) blown onto the slide were trapped in the vaseline. When a box of slides, about two dozen, were accumulated at each station, they were shipped to St. Paul, and a portion of each slide, usually about a square centimeter, was examined microscopically and the number of urediospores of the stem rust fungus was tabulated. These spores are characteristic enough in size, shape, and color so that one familiar with them and with related kinds can identify them with certainty. This meant that someone had to spend most of his working hours for a couple of months ''reading'' the slides so exposed, a tedious but necessary job.

Some of the later students of aerobiology, who used more sophisticated, or at least more complicated, methods and devices to collect spores, objected to this ''gravity impact'' or wind-impact method of collecting spores, on the grounds that it did not accurately determine the quantity of spores in a given volume of air. True. But it probably was much closer to actual plant exposure — where the spores either fall or are blown onto the leaves of the wheat plants in the field — than were the more complex volumetric devices. In this survey of rust spores in the air the important thing was not the number of spores per cubic foot or cubic meter of air but the number deposited in a given period of time per unit of surface area.

The figures given by Stakman, above, indicate what a tremendous area of land can be effectively inoculated by a single spore shower. As it happened, in 1925 rains followed this spore shower in much of the northern wheat-growing area, and a disastrous epidemic of black stem rust resulted.

In the fall, with strong north and northwest winds, spores of this same rust are blown south from Canada and the Dakotas to Texas, to infect the young winter wheat plants there. In other words, these urediospores of the stem rust fungus, carried passively by the wind, follow much the same route as the migrating birds in that region. All they have to do is stay airborne and they will be carried from northern Mexico and southern Texas up through the Dakotas and into northern Manitoba and the other Prairie Provinces of Canada in the spring, and back down in the fall. Pady and Kapica (69) reported that fungus spores were relatively scarce in arctic air masses, and relatively common in tropical air masses, over the Atlantic between Montreal, Canada, and London, England.

Puccinia graminis tritici is just one rust fungus the spores of which are very numerous in the air at many times and places. There are more than 4000 species of rust fungi that parasitize many different kinds of wild and cultivated plants, including common plants in our yards and gardens. Spores of some of these must be relatively abundant in the air, and just about everyone must at times inhale a good many of them; yet these spores do not seem to be much involved in respiratory allergy or in other allergies, even as some pollens, notably those of pine and other conifer trees, which sometimes are very abundant in the air, are not responsible for allergies.

One of the first allergists to attribute respiratory allergy to inhalation of fungus spores was a Canadian who, in 1924, reported that three men came down with an attack of asthma when they were exposed to a heavy concentration of rust spores at threshing time — most probably the urediospores of the black stem rust fungus. However, Feinberg (60) questions whether the allergic response in these men was caused by the rust fungus; he suspected that it was due to spores of other fungi that accompanied the rust fungus. In support of this contention he adduced the fact that he never had observed any reaction in patients given skin tests with extracts of pure rust spores but frequently did observe reactions in such tests to extracts from rust spores contaminated with *Alternaria*, *Hormodendrum*, and *Penicillium*. Feinberg from his own observations knew that spores of rust collected in the field were likely to be contaminated with spores of these other known allergenic fungi and that "pure" rust spores were hard to come by — Feinberg was a very astute investigator who took very little for granted.

Fungi other than those described above may produce large numbers of spores also, and many of these spores become airborne. Durham (59) mentions a spore shower in early October 1937 in which thousands of tons of spores of *Alternaria* and *Hormodendrum*, originating presumably on stubble and other plant remains in the fields of Minnesota and the Dakotas, were carried eastward and southeastward for hundreds of miles, also presumably out to sea and perhaps to Europe. "Thousands of tons" is an almost incredible amount of spores, but spores of these two genera of fungi are present in the air in some numbers just about everywhere at just about all times where the spore content of air has been monitored, which may be one reason why they are important in respiratory allergy, about which more later on.

Methods of Sampling Number and Kinds of Fungus Spores in the Air

Once the allergists recognized that airborne fungus spores might be an important contributing cause of respiratory allergy, they and others began sampling the air to determine the number and kinds of fungus spores present in the air, and for a couple of decades such sampling was carried on extensively in many places.

Basically there are only two or three methods to determine the number and kinds of fungus spores present in the air: (1) The spores are caught, by gravity or by an artificially generated current of air, on a glass microscope slide (or on a bit of transparent cellophane tape) that later is examined microscopically, and the number and kinds of spores deposited on a given area are recorded. (2) The spores are deposited on a nutrient agar medium, where they germinate and grow into colonies, and these colonies are counted and identified as to genus. (3) In a sort of combination of (1) and (2), the air is drawn through water or through a suspension medium in a small container, by means of a suction device, and the spores and other material in the air are washed out by the liquid; the liquid can then be mounted on a slide and observed microscopically and can also be plated on an agar medium or on a variety of agar media, and the resulting colonies can be counted and identified.

In the early days of aerobiology, up to the 1940s, the collection of spores was done in one of two ways: either slides coated with vaseline were exposed, usually for 24 hours, on a wind vane that kept them pointed into the wind, or on a platform of some sort, and then these were taken to the laboratory and examined at leisure; or petri dishes containing an agar medium were exposed, usually for 2 to 30 minutes, with the dishes then taken to the laboratory and incubated until the colonies could be counted and identified. Since the early 1940s more complex devices have been developed that deposit airborne particles from a known volume of air either on slides or on agar in petri dishes.

Each of these methods — impaction of airborne material on a microscope slide, for later identification of the deposited spores, or impaction on an agar medium with identification of the resulting colonies — has its own limitations and virtues. Positive identification of spores deposited on slides, or on transparent tape, is limited to those kinds of spores having characteristic size, shape, or ornamentation, which does not include a very great number of species or genera of fungi whose spores are likely to be present in some numbers in the air at many times and places. Also a great amount and variety of other material is likely to be deposited on slides that are exposed out-of-doors — dust and dirt, fragments of plants and insects, minute barbules from birds' feathers, and all sorts of biological and nonbiological junk and debris commonly carried by the air. Some experience and expertise are required on the part of the examiner to detect the particular kinds of spores in which he is interested. If the main object is to detect the spores of a certain species of rust or smut fungus, or of *Alternaria* or *Hormodendrum*, most spores of which are identifiable with a considerable degree of assurance, the method is very useful — and it is for the detection of these fungi that the method has been mostly used. In the hands of an expert and experienced mycologist and aerobiologist such as Hirst (64), who used a device that impacted spores onto a slide moving at a known rate over a 24-hour period, this technique can yield some surprising information about what kinds of spores prevail in the air near the ground at different times of the day and night and under different weather conditions.

The identification of fungus spores collected on agar, by whatever means, and allowed to develop into identifiable colonies, is limited to those that can develop on whatever culture medium is used. Special culture media have been developed for individual groups of fungi, such as *Fusarium*, or for storage fungi, and are reasonably effective in detecting those fungi for which they were designed. The technique will not detect spores or other propagules of obligate parasites, or mushrooms, or wood-rotting fungi, or a host of other fungi.

It seems reasonable that the value of whatever method is used be judged by how well it fulfills the purpose of detecting whatever it was designed to detect, not on what it fails to detect.

So much for the techniques. Anyone interested in following this aspect of the problem further is referred to the papers already cited by Ingold and Hirst, plus one by duBuy and Lackey (57) dealing with comparisons of different sampling methods and devices. So far as respiratory allergy is concerned, it appears the *Alternaria*, *Hormodendrum* (*Cladosporium*), and a few of their relatives cause most of the trouble, and these are detected at least moderately well by even the simple exposure of slides or of petri dishes containing agar.

Fungus Flora of Air Outdoors

Most surveys of numbers and kinds of fungus spores in the air have been made outdoors, which is reasonable because that's where most of the problems are, both in epidemiology of plant diseases and in the fungi involved in respiratory allergy. There are exceptions to this, naturally, and some of these will be taken up later. Depending on the objectives of the work, or on convenience, or both, this sampling has been carried out from near the ground level, where most plants and most people spend most of their time, to as high as the sampling devices could be carried by plane or balloon. No such surveys have been carried on in outer space, beyond the confines of the atmosphere, although it would be interesting to find out what, if anything, in the way of fungus spores and bacteria might be floating around up there.

Heights at Which Fungus Spores Have Been Detected

In the 1920s and early 1930s spores of the wheat stem rust fungus were caught by several workers above 10,000 feet — usually simply by holding a greased slide out of an open airplane cockpit by hand or on a special holder. No spores were reported above about 14,000 feet simply because that was as high as the planes used in this work could go, or as high as the spore-trapping plant pathologist wanted to go without auxiliary oxygen. When a graduate student on the project returned from such a flight he was the envy of his fellows and would regale them interminably with lies about his experiences at that fabulous altitude.

In 1935 the balloon *Explorer II*, under the auspices of the National

Geographic Society and the Army, carried among other things a spore trap that was released at an altitude of 71,395 feet and that was set to close (and presumably did close) at 36,000 feet. During its fall it sampled an estimated 2380 cubic feet of air — but pretty thin air, of course. Living spores of five common molds were recovered from the trap, proof that the spores of these fungi, and, presumably, those of many other fungi, could survive whatever hazards they were exposed to in being carried to that altitude.

Distances to Which Spores Can Be Carried

It already has been mentioned that urediospores of the black stem rust fungus in 1925 were carried in tremendous numbers from Oklahoma and southern Kansas to the Dakotas, some hundreds of miles away. Canadian investigators caught spores of the same fungus in the Peace River valley, at least 700 to 800 miles from the nearest possible source. In the 1930s, Charles Lindbergh, in collaboration with the United States Department of Agriculture, exposed slides periodically from his plane when flying between Maine and Denmark (he was surveying the Great Circle Route for commercial flights to Europe) and caught spores of a "considerable number" of molds above the Arctic Circle and, at altitudes of 3000 feet, over the open ocean far from land. The balloon *Explorer II* encountered winds of 40 to 60 miles per hour at the heights at which spores were trapped, and it was calculated that the spores at that altitude and with that wind could in a week be carried 8400 miles. More about this sort of calculation shortly. The 300-miles-per-hour west to east jet stream that is encountered around 35,000 feet, usually just when dinner is being served on a plane, was not known then. Using the same simple calculation as was used above, we can see that, with its aid, in a week of 168 hours the spores could be carried 50,400 miles, or at that 45-degree latitude about three times around the world. So far as I am aware, the microbiological content of the air in this jet stream never has been investigated; it would be extremely interesting to do so, and not especially difficult either. We still do not know to what altitudes fungus spores and bacteria can be carried and how long they can persist at high altitudes and still germinate and grow.

That the air, just about wherever it has been sampled, carries a fairly large and varied mycological freight is not at all surprising. We long have known that moving air has a great carrying capacity — it redistributes farms and farmers, and fine material may remain suspended in it almost indefinitely; the volcanic dust that became airborne when Krakatoa in the Sunda Strait in Indonesia blew its top in 1883 made red sunsets around the world for a couple of years. The wind is a very effective dissemination agent of the spores of a great number of fungi.

Airborne Insects

For a long time entomologists in general have been well aware that many kinds of nonflying insects have special adaptations for being effectively carried by the wind — witness the small spiders that on a quiet summer day sometimes are seen blowing past by the thousands, each at the end of a short thread. Darwin mentions a "shower" of such airborne spiders off the coast of South America, fairly well out to sea — in that case they may have picked the wrong wind. Glick (61) made 1358 flights between August 1926 and October 1931, most of them over Louisiana but some over Mexico, to collect airborne insects. He says, "Numbers of adults, nymphs, and larvae of wingless forms of insects and mites were collected in the upper air at altitudes as high as 14,000 feet and a spider at 15,000 feet. These wingless forms are all at the complete mercy of the upper air currents." When the air was smooth most of the airborne insects were taken at an altitude of only a few hundred feet, but in rough and turbulent air, with ascending currents, the insects were most numerous at altitudes up to 5000 feet. He also caught seeds of a number of different families of plants, usually in turbulent air. He did not say whether these seeds were alive, but there is no reason to suppose that some of them were not; even very small seeds of many kinds of plants are remarkably tough and durable.

Rate of Fall of Fungus Spores in Still Air

One aspect of fungus spore dissemination with which aerobiologists occupied themselves for some time was that of determining the rate of fall of spores of different kinds and different sizes in still air. Why this was considered vital to or even supportive of the fact of long-distance aerial dissemination of fungus spores now seems obscure. If spores are carried to altitudes of several miles and are caught hundreds of miles away from any possible source — both of which facts had been well established by the early 1930s — then they obviously can be airborne for some distance. If we suffer a heavy fallout of red dust in Minnesota, as we sometimes do, and the meteorologists tell us that the wind came more or less directly northeastward from Oklahoma, and the soils people tell us that the red dust was picked up in Oklahoma (they saw the wind picking it up in dense clouds there, and others reported these clouds traveling northeastward), we do not need data from tests on the rate of fall of dust particles in still air to convince ourselves that dust can be carried for some distance by the air.

It also has long been known that when slides are exposed horizontally in outdoor air, with both the upper and lower sides exposed to the air, about one-half as many spores are caught on the under or lower side of the slide as on the upper. In air so still that one can see clouds of spores liberated from a fruit body of *Fomes applanatus*, the cloud does not move downward, but drifts off at an upward angle before it thins out and disappears. As mentioned earlier, the tops of fruit bodies of this fungus usually are covered with a deposit of brown spores liberated from the underside of the fruit body — they ''fell'' upward. Nevertheless, the rate of fall of spores was studied by a number of people, and representative results are given in table 5-1. Some of these tests were made by no other than C. M. Christensen; I am sure that at the time I was confident that I was making a Contribution to Science. Well, at least, it was rather interesting.

Once these fungus spores reach a height of more than a few inches, and certainly when they reach the height of a few miles, they probably never do ''fall'' out of the air, but are borne along indefinitely until they are carried down by air currents or by rain or snow — the raindrops at the beginning of a shower usually contain large numbers of fungus spores. From the standpoint of epidemiology of plant diseases, the important thing is not only how far the spores that cause rusts and smuts and blasts and blights and mildews are or can be carried, but also whether they arrive in living condition and at a time when their host plants are susceptible to infection and when the weather is favorable for such infection. That combination of circumstances may happen rather

Table 5-1. Approximate Rate of Fall of Spores of Plant Pathogens in Still Air, and Their Approximate Theoretical Dispersal Distance

Organism	Size of Spores in μ	Rate of Fall per Second in Millimeters	Time Required to Fall 1 Foot	Time Required to Fall 100 Feet	Dispersal Distance in 20 MPH Wind and at 1 Mile Altitude (in Miles)
Alternaria sp.	20 × 10	3	1 min. 40 sec.	2¾ hrs.	2,900
Cronartium ribicola	22 × 19	8	38 sec.	1½ hrs.	1,580
Helminthosporium sativum	75 × 20	20	15 sec.	25 min.	440
Puccinia graminis tritici	28 × 17	12	25 sec.	42 min.	740
Puccinia triticina	25 × 20	13	25 sec.	39 min.	680
Ustilago zeae	10 × 9	3.5	1 min. 26 sec.	2-2/5 hrs.	2,500

frequently in modern agricultural countries, where thousands of square miles of land are covered with a given variety, or with a few closely related varieties, of crop plants, all susceptible to infection and destruction by a single species or a single race of parasitic fungus. This happened with oats in the 1940s. Plant breeders had produced a number of varieties of oats, all from the same two parents, that monopolized the oats acreage in the northern United States. They were resistant to rusts and smuts, and they yielded as much as 50 percent more than the varieties previously grown. A fungus unknown up to that time, soon to be named *Helminthosporium victoriae* (one of the parents of these new hybrids was named Victoria), began to infect and cause blight on the new varieties, and in 1946, 1947, and 1948 almost ruined the oat crop throughout the major oat-growing states. The same thing happened with corn in 1971, when *Helminthosporium maydis*, which causes southern leaf blight of corn, ruined much of the corn crop from the Gulf Coast to the northern corn belt; practically all the corn grown throughout this huge region carried the TMS (Texas male sterile) gene that included susceptibility to this fungus. That host uniformity must occur very seldom in the wild, and so for survival in the wild those plant-parasitic fungi whose spores are disseminated by the wind had to produce relatively large numbers of spores, and they also had to develop spores that would endure drying, freezing, and exposure to ultraviolet radiation for some time without losing their viability. Spores do not have to be alive to cause respiratory allergy, of course; that they do cause such allergy is just one of the accidents of evolution; so far as the fungi are concerned this allergenic effect has no conceivable survival value.

Fungus Spores within Buildings

Some people may be without respiratory symptoms when outdoors but develop a severe case of sniffles and sneezes and asthma when sweeping or vacuuming or otherwise shifting around the dust within the home, or when making beds, or when otherwise occupied in the home. Everyone must at times have seen, when the sun was shining in through the bedroom or living room window at just the right angle, an abundance of fine floating material in the air. Among the particles are sure to be some spores of fungi, and in some houses this impalpable and almost invisible dust contains many spores of fungi, including some that may be peculiar to a given house.

Samples of house dust were collected from vacuum cleaners in 76 homes in St. Paul and Minneapolis during the winter months. In that region in the winter the wind, blowing mostly from the polar regions and crossing hundreds of miles of snow-covered prairies and forest, does not carry a heavy load of fungus spores; the doors and windows of houses are not open much; and there is a minimum chance of contamination of indoor air with spores from the outside. The samples of house dust so collected were diluted in a sterile suspension medium and plated on various kinds of agar in petri dishes to determine the number and kinds of fungi present; the numbers of bacteria that would develop on a single medium also were determined.

The number of colonies of fungi got from one gram (about a teaspoonful) of house dust averaged 179,666 and ranged from a low of 6000 to a high of 3,200,000. Bacteria per gram of house dust averaged 10,700,000 and ranged from a low of 1,144,000 to 20,000,000.

Petri dishes containing an agar medium were also exposed in these homes, then were incubated to see what fungi would grow. The numbers of colonies of culturable fungi, as determined by this method, were lower than those usually found in outside air in late summer and fall, when fungus spores are likely to be most numerous in the air, but they were mostly of different kinds from those encountered in outside air — chiefly *Aspergillus* and *Penicillium*. The same two genera also made up the bulk of the colonies of fungi recovered from the samples of house dust. The greatest numbers of spores in the air within homes were found just after sweeping or vacuuming or dusting, all of which activities transfer into the air some of the dust from wherever it has settled.

In many of the samples of house dust, by far the most prevalent fungus was *Aspergillus restrictus*, followed by *A. glaucus*. Both of these species of fungi are xerophytic and are capable of growing in a great variety of materials whose moisture contents are in equilibrium with relative humidities of 68 to 75 percent; that environment, in fact, constitutes the ecological niche in which they can predominate over all competitors. The critical thing for their growth is not the makeup of the substrate, but rather the relative humidity or moisture content, or osmotic pressure.

No one would deliberately maintain a relative humidity of 68 to 75 percent in the air within a home in the winter — that is, in the air flowing past the humidity gauge. However, air of 35 percent relative humidity does not have to be cooled too much to reach a relative humidity of 70 to 75 percent, as witness the condensation of moisture on windowpanes in the winter, some of which moisture will run down and accumulate in the paint and wood, which promptly become moldy. In summer, in all but the driest regions of the world, the air within homes often is of high enough relative humidity to permit fungi to grow in many things, especially in cool basements, and enough spores may be produced by the fungi in those humid periods to contaminate the air throughout the home fairly heavily for the rest of the year. In many homes even in the northern and relatively dry regions of the United States there is condensation or leakage so that wood will decay, and this is a considerably higher moisture content than is required by the xerophytic fungi such as *Aspergillus restrictus* and *A. glaucus*.

Another ample source of moisture, especially for items like furniture stuffing, mattresses, and pillows, all of which are common sources of allergenic fungi in homes, motels, and hotels, is that given off by our own bodies. We do not have to perspire or sweat visibly — we are constantly evaporating water from all of our body surface. If the air is so humid that this cannot evaporate as fast as it is produced, it accumulates as perspiration. If the air is dry, this water is given off as vapor, but there is no less of it given off then than when we perspire or sweat — it just is not visible. If you sit in an overstuffed chair some of this moisture will accumulate in the stuffing, and if you lie on a mattress, with your head on a pillow, some of this moisture will accumulate in the mattress and in the pillow stuffing, and these almost inevitably in time become moldy, some of them heavily so. And the major fungi that grow in these household goods are not the high-moisture-requiring kinds such as *Rhizopus* or *Fusarium* or *Alternaria*, which some mycologically innocent allergists have isolated from such materials, but *Aspergillus restrictus* and *A. glaucus*. Some strains or races of these two species will not grow on agar media unless the media contain from

6 to 10 percent or more of salt — sodium chloride — and so they are not likely to be detected in routine sampling of such materials by allergists, or even by mycologists who are unaware of their presence and of their requirements for growth. This may not matter too much, since it is claimed that different species of a given genus do not differ appreciably in their ability to elicit an allergic response, so if a person is not sensitive to *A. niger*, a fungus with which all allergists are familiar, and for sensitivity to which they almost certainly will test their patients if they have reason to suspect mold-caused allergy, he should not be sensitive to *A. restrictus* or to *A. glaucus* either. However, physiologically and ecologically, *A. restrictus* is about as different from *A. niger* as pines are from pineapples. It would be interesting to see some tests made, in patients sensitive to house dust or mattress or pillow stuffing, with *A. restrictus* and *A. repens*, often the major fungi present, rather than with *A. niger*, which seldom is present in any numbers. This would give us actual experimental evidence on the possible involvement of *A. restrictus* in respiratory allergy, and especially in allergies to house dust and mattress stuffing. Now we have only logic, and evidence is almost always preferable to logic, at least in establishing truth, or probability, in biology. Foam rubber cushions and mattresses and pillows have often been advertised as nonallergenic and mold-free. The tests on which the "mold-free" statement is based were made with foam rubber of so high a moisture content that xerophytic fungi could not grow. If foam rubber is kept at 75 percent relative humidity, *A. restrictus* and *A. glaucus* will grow in it and produce spores; they will not grow lushly, or produce a heavy crop of spores, but they will grow and sporulate.

An interesting case will illustrate what may happen in this sort of sensitivity. Allergists in Johannesburg, South Africa, encountered a patient who had what at first appeared to be a rather bizarre type of respiratory allergy — "seaside allergy." He was a young man of good health who regularly and inevitably came down with severe bronchial asthma when he visited the seaside, where he took groups of boys on camping trips. The attacks occurred mainly or almost only at night. They were mild at first, but became progressively worse from night to night, until finally he could scarcely walk or eat, and could not sleep at all. Soon after he returned home from the seaside the symptoms subsided, and did not recur until the next camping trip. This had been going on for some ten years before he consulted allergists. Obviously there was something in the seaside air, and especially at night, and most probably in his tent, to which he had become sensitized.

On the basis of the history of the case, its time and place of occurrence, plus skin sensitivity tests, the allergists he consulted more or less ruled out any food or pollen allergies. The man slept in a tent, and during the noncamping season the tents and mattresses were stored in an outhouse, away from the main house, and the allergists thought that either the tent material itself might be involved or the mattress might be moldy. Tests were performed that allowed the tent material to be excluded as a possible cause. Then the allergists examined and plated out samples of material from the mattress and recovered several genera of fungi from it — *Mucor, Penicillium*, and *Fusarium*. Now it is highly unlikely that *Mucor* and *Fusarium* were growing in the mattress stuffing — it would have had to be practically soaking wet for them to do so since they require free water to grow, and no camper in his right mind would sleep on a mattress that wet. Had they cultured the material on a medium that would reveal the xerophytic species of *Aspergillus* they probably would have found them to be present in large numbers. However, they grew the fungi that they had isolated from the suspect mattress, made a combined extract, injected a bit of this into the young man's skin (a standard procedure to test sensitivity), and elicited a strong positive reaction. The patient was advised not to sleep on a moldy mattress. On the next camping trip to the seaside he slept on the moldy mattress the first night because of foul weather, and developed bronchial asthma. After that he stayed off the moldy mattress and remained free of symptoms. Of course he probably could have obtained relief also simply by buying a new and, one would hope, clean mattress — although some lots of cotton and kapok are pretty moldy before they are even made up into mattresses or furniture stuffing.

A rather extreme case of a home being taken over by mold was described in a local newspaper a few years ago. It may be somewhat exaggerated, but on the other hand there must be some truth in it because it seems unlikely that a reporter would know enough about fungi to make it up. According to this account, a Baptist minister and his family were driven from their home, in the foothills of the Blue Ridge Mountains in North Carolina, by a mold. The house was an old one, his ancestral home, as a matter of fact — he had been born there some 60 years before. He described the mold as a "sour and fiery sort of thing"; it took over the entire house, from cellar to attic, covering the walls with powdery spores, growing over the floors and furniture, and rotting the clothing in the closets, where the masses of spores rose in tiny puffball shapes. Attempts to control it by scrubbing with disinfectants proved futile. The minister's wife and daughter were allergic to it, their eyes watered, they could not breathe, their sleep was fitful. The family abandoned the house, taking with them only a stove and refrigerator, metal things that could not be invaded by the loathesome mold.

Assuming that the account is basically true, there may be a rational explanation for the strange occurrence. That region is warm and humid. Houses and other buildings constructed there of decay-susceptible wood are very subject to invasion by wood-rotting fungi. If constructed in part of decay-resistant wood such as white oak heartwood or cypress heartwood they will be decayed more slowly. At the present time houses built in such a region without recognition of this hazard may, within a year or two, have their floors decayed to the point where someone walking across the room literally falls through the floor. The old house of the newspaper account may have been thoroughly invaded by this "dry rot" — which is a misnomer because the fungi that cause it require as much water as any other wood-rotting fungi (in wood this means a moisture content of about 30 percent as a minimum). If the walls of the house were completely decayed by fungi, the wood might have been moist enough (partly decayed wood often is hygroscopic) in that humid climate to support a heavy growth of *Penicillium* or *Aspergillus*. In this case washing the surface of the walls, even with a disinfectant, would not do much good — the fungus would simply produce another crop of spores in a day or so. The family probably were lucky that they moved out when they did, before the house collapsed around them.

In England and in northern Europe dry rot caused mainly by the fungus *Merulius lachrymans* is relatively common. Mycologists interested in airborne spores sampled the air within two houses in which this fungus had caused decay and in which it had produced fruit bodies. Both houses had been abandoned, probably in part because of the progressing decay which was impossible to live with and impossible to arrest. They used a motor-driven sampler called a Cascade Impactor, which deposits on slides spores and other airborne particles from a measured quantity of air. The spores of *M. lachrymans* can be recognized with assurance by anyone familiar with them, as these men were, and there were few other spores and little other airborne material in these abandoned but closed-up houses.

In the first house so sampled, there were spore-producing fruit bodies of *Merulius lachrymans* in the cellar. There, in the cellar, 80,000 spores of *M. lachrymans* were caught per cubic meter of air; on the ground floor there were 26,000 spores per cubic meter of air and, on the next floor up, 16,000 spores per cubic meter. All these spores came from the fruit bodies in the cellar and were carried throughout the closed and empty house by air currents.

The second house was 400 years old and had once been a country mansion, but now was inhabited only by the dry-rot fungus and maybe by assorted haunts. In the cellar, where again there were fruit bodies of *Merulius lachrymans*, they caught 360,000 spores of this fungus per cubic meter of air, and two floors above they caught 1360 spores per cubic meter of air. They say in their report (63), "Evidently spores can travel in slow convection currents from rooms and cavities containing a sporulating fruit body to distant parts of the house." Indeed they can, as tests in the plant pathology building on the St. Paul campus of the University of Minnesota had already shown. These tests were interesting enough to justify a summary account here.

The fungus used was *Hormodendrum resinae* (it should have been called *Cladosporium resinae*). In nature this fungus is found almost only in resin-impregnated soil or in wood impregnated with coal tar creosote to protect the wood from decay. The wood treated with creosote is well protected from decay, but not from invasion by *H. resinae*. The fungus will grow readily on an agar medium that contains enough creosote to prevent the growth of any and all other fungi. Soil and decaying plant materials heavily overgrown with a multitude of common and uncommon fungi when scattered on such agar will not yield any colonies of fungi — evidence not only that fungi other than *H. resinae* will not grow on this medium, but also that *H. resinae* is not present in such soil or decaying plant remains. In repeated tests in which creosote agar and creosoted wood blocks were exposed to outdoor air and then incubated, no colonies of *H. resinae* ever developed, evidence that its spores are not common in outdoor air. Spores of this fungus must occasionally be present in outdoor air, of course (probably especially around creosote treating plants), since its spores are airborne, but at least it never has been caught from outside air. So in the creosote agar we have a highly selective medium in which no other fungi will grow, and in *H. resinae* we have a marker fungus that normally is not encountered in the air either indoors or out, but which will grow readily on the creosote agar.

When grown on malt agar, *Hormodendrum resinae* produces an abundance of powdery spores that can be readily brushed into the air. The spores are only a few microns long, and in still air in a closed tube they fall only about an inch per minute, and so are suited to transport by even slow-moving air currents. They retain their viability for days or weeks at the relative humidities ordinarily encountered in occupied buildings. So far as is known, this fungus in innocuous to man, beast, or plants, and so the presence of considerable numbers of its spores in the air for a short period of time should result in no harm or inconvenience.

Spores of *Hormodendrum resinae* were brushed off the surface of an agar culture in a room on the first floor of the plant pathology building, a four-story structure with a stairway at each end and a hall down the middle of each floor. The building has no forced-air circulating system. Before the spores were liberated, dishes of creosote agar were exposed for a half hour at various stations throughout the building; then the dishes were covered and stored in the laboratory. These served as controls; no single colony of *H. resinae* (or of any other fungus) ever developed on any of them. Another set of dishes were exposed at the same stations on each floor, and the spores were liberated. Still another set of dishes were exposed on each floor at intervals of five minutes or more, the preceding ones being stacked in the laboratory and left until the colonies of *H. resinae* developed on them and could be counted. A number of tests were made, and representative results are summarized in tables 5-2 and 5-3.

As can be seen from the tables, within a few minutes after the spores were liberated in a room on the first floor, they were caught in relatively large numbers throughout all four floors; they permeated the air of the entire building and some even were caught in rooms of which the doors were closed throughout the test period.

There is a famous picture of a sneeze that shows more than 40,000 droplets being expelled from the sneezer's mouth. Judging from the results of the dissemination of spores of *Hormodendrum resinae*, someone four or more floors below us can sneeze in our face rather effectively. Subsequent tests of the same sort showed that spores traveled from the upper floor to the lower floor just about as fast as they traveled from the lower to the upper floor of the plant pathology building. This sort of intramural dissemination can be important in hospitals, if the particles disseminated are infective, and in some kinds of food-processing plants where it may result in contamination by highly undesirable fungi and bacteria.

The oyster mushroom, *Pleurotus ostreatus*, is a common gilled fungus that grows on dying and dead hardwood trees, the wood of which it decays. A fresh fruit body of this mushroom brought into the laboratory and properly set up will shed a visible cloud of spores for 24 hours or more — we often use it as a class demonstration. One rather small fruit body, only a couple of inches across, was brought into the laboratory and left on the table for half an hour. The room was 20 by 20 by 12 feet high, with a volume of 4800 cubic feet. Glass microscope slides were placed at various locations in the room, on tables and on shelves as high as seven feet. After half an hour the fruit body was removed and the slides were examined. There were on the average eight spores — with almost no variation above or below this figure — per high power field of the microscope in every field examined on every slide. The area of the high power field in the microscope used for this examination was just about equal to the dot over the i in this printing. There was no fan or other device in the room to provide artificial air currents — the natural convection currents always present distributed the spores uniformly throughout the room.

Table 5-2. Number of Colonies of *Hormodendrum Resinae* on Plates of Creosote Agar Exposed in Successive Five-Minute Periods on the Second, Third, and Fourth Floors after Spores Were Liberated in the First-Floor Hallway

Floor	Number of Colonies per Dish			
	First 5 Minutes	Second 5 Minutes	Third 5 Minutes	Fourth 5 Minutes
2	85	70	27	20
3	20	33	9	16
4	4	17	10	7

Table 5-3. Number of Colonies of *Hormodendrum Resinae* on Plates of Creosote Agar Exposed for Different Periods of Time after Liberating Spores in a Room on the First Floor

Location of Exposed Plates	Number of Colonies per Dish Exposed for Given Period of Minutes							
	0–5	5–10	10–20	20–30	30–60	60–120	120–240	Total
First-floor hall	228	49	43	31	26	15	5	397
Second-floor room, door closed	0	0	1	1	3	10	2	17
Third-floor greenhouse connected to the main building by a 30-foot passage	0	0	0	0	1	0	1	2
Third-floor laboratory, door open	8	0	5	19	55	28	8	123
Third-floor hall	0	8	70	35	44	12	0	169
Fourth-floor hall	68	78	50	34	51	7	3	291
Fourth-floor room, door open	0	0	2	3	7	14	5	31
Total	304	135	171	123	187	86	24	1,030

Airborne Spores in Grain Elevators

There are a number of industries or industrial processes in which the workers are likely to be exposed constantly or intermittently to air with a high dust content, of which fungus spores make up at least a considerable part. One of these is grain handling, especially after the grain has been in storage for some time and some of it has become invaded by storage fungi. Dilution cultures of dust collected from grain elevators have yielded up to several million colonies per gram of dust, and especially of those species of *Aspergillus* that make up the main fungus flora of stored grains. At some times and places in some elevators when grain is being transferred, the dust is so thick that one cannot see more than 50 feet or so through it. Grain elevator operators uniformly and stoutly maintain that they keep their grain at moisture contents too low for fungi to grow, yet when petri dishes containing an agar medium developed especially to reveal storage fungi were exposed, even for so short a time as one-half minute, in such an elevator, and then were incubated for a time, hundreds of colonies of *Aspergillus restrictus*, *A. glaucus*, and *A. candidus* developed on them. Similar dishes, exposed in grain fields near the elevator, and for as long as half an hour, and similarly incubated, yielded no colonies. The conclusion is that the inoculum of fungi caught from the air within the elevator were coming from fungi growing on the grain and dust and debris within the elevator. Another reasonable conclusion is that elevator operators do not know as precisely as they should the moisture content of the grains they store. ''Elevator operators' lung'' is a recognized occupational hazard. Those who work in and around grain elevators should wear dust masks, and some of them at times do so.

Airborne Spores in Greenhouses

The air in greenhouses may at times bear a heavy load of spores of given kinds of fungi, as probably would be expected. Greenhouses in which plant pathologists work are likely to contain plants heavily infected with rusts, smuts, blights, and mildews — that's what the greenhouse is there for; yet this does not appear to constitute any special occupational hazard to plant pathologists, at least I never have personally known or heard of greenhouse allergy among plant pathologists.

Respiratory Allergy Caused by Airborne Fungus Spores

Until about 1930 only a few cases of respiratory allergy were attributed to exposure to airborne fungus spores; allergists in general did not consider mold-caused allergy to be of any great importance, and some even questioned whether it occurred at all. Evidently this attitude stemmed from the fact that they had not been taught that mold-induced allergy was a problem and had not thought of it themselves; therefore it could not be, an attitude by no means restricted to allergists.

In the 1930s, however, several allergists began to investigate the possibility that some cases of respiratory allergy might result from inhalation of fungus spores. They suspected this in part because they found many fungus spores on slides exposed to catch pollen grains (they had to be able to recognize the fungus spores among all the other stuff on the slides, so they must already have put in some time studying fungi), and in part because some of their patients had outbreaks of hay fever and asthma when few pollen grains were present in the air. Their published case histories soon convinced others that this was worth looking into, and it was not long before surveys of airborne mold spores were under way in many different places. Dr. Feinberg, a prominent allergist in Chicago, who was one of the first to establish the fact that airborne fungus spores could cause respiratory allergy in humans, exposed culture dishes daily for five years and identified the major fungi so detected — 52,000 colonies on 3652 plates. Of these fungi, *Hormodendrum* made up 42 percent, *Alternaria* 30 percent, *Penicillium* 11 percent, and *Aspergillus* 4 percent. Miscellaneous fungi, of which he identified 13 genera, and unknowns made up 13 percent. He recognized that with the techniques he used, spores of many fungi present in the air went undetected. He also knew by that time from his extensive clinical experience that most of his patients with respiratory allergy were sensitive to *Alternaria, Hormodendrum*, and a few of their relatives in the same fungus family (*Dematiaceae*), so he was detecting the major culprits among the fungi.

Subsequently a collaborator had slides exposed in summer and fall, the peak times for fungus spores in outdoor air, in 92 cities in 40 states of the United States, in Alaska (then a territory), and in two foreign countries. He does not say in his report (59) whether a slide was exposed daily in each location, but this was the usual practice at the time. If this was done, then the number of slides exposed and subsequently examined microscopically (one square centimeter per slide) for the presence of fungus spores must have been in the thousands. He identified only the two most common spores, *Alternaria* and *Hormodendrum*. The largest number of spores, 2820 of *Alternaria* and 1820 of *Hormodendrum*, were caught at Moorhead, Minnesota, opposite Fargo, North Dakota, in the Red River Valley, and the lowest number — zero counts of both genera — were at Juneau and Nome, Alaska. Stations in several midwestern states had consistently higher counts than other regions, including even stations in the deep South where high temperatures and high humidity combine for what would seem to be ideal conditions for production of spores by these two genera of fungi. Presumably a major source of the spores of these fungi in the Midwest was straw and stubble in the grain fields; in a moderately moist harvest season the stems and leaves of these plants over thousands of square miles may be almost black with the spores of these fungi.

From then on, spore trapping was carried on in many locations in the United States and in other countries. In general the patterns were more or less similar — spores of a few kinds of fungi, and usually much the same kinds, made up the great bulk of the airborne spores, and the times of peak appearance of these spores in the air was late summer and fall. This sort of sampling has not been discontinued, but it is much less extensive now than it was in the 1940s and 1950s; now it is used mostly not to detect spores of the principal airborne fungi such as *Alternaria* or *Hormodendrum*, but for tracing the spread of specific plant pathogenic fungi such as peanut rust or coffee rust, or for tracing the spread of specific pathogenic or allergenic agents.

6

Fungus Predators and Parasites of Nematodes and Insects

Among the ecological niches explored and occupied by some of the fungi are the smaller animals, especially nematodes and insects, in the populations of which they help to maintain the balance of nature. This is not a steady and static balance, but a shifting and fluctuating one involving a complex interplay of many different kinds of living organisms, all of them influencing one another and in turn being influenced by such nonliving forces in the environment as temperature and moisture, the physical and chemical makeup of the soil or other substrate in which they live, electrical fields, gravity, phases of the moon, and probably others that we still know nothing about. We often try to shift this balance in our favor by trying to hold in check those fungi that attack the beneficial insects such as the silkworm and the honeybee, or by increasing and spreading around in the environment those fungi that trap and kill plant-parasitic nematodes or that infect and kill noxious insects. This chapter aims to summarize some of the information on these almost ubiquitous and always interesting fungi that almost never are seen except by those few specialists who happen to study them.

Fungi as Predators of Nematodes

We shall look first at fungi that prey on nematodes; to understand what goes on in that strange and sometimes implausible world it is necessary to know something about what nematodes are and what they do.

Nature of Nematodes

If relatively few people have seen nematodes it is only because they do not know how and where to look for them (assuming that they might be interested in seeing them in the first place) because everyone from infancy on has frequent and intimate contact with many kinds of nematodes, some good, some bad. The name nematode means thread-like, which is descriptive enough as scientific names go. "Eel worms" is another name for them, and the common vinegar eel, long used in biology classes because it is so easy to come by and so easy to grow, is a fairly representative example of a free-living nematode.

Nematodes have a comparatively simple structure, consisting of a muscle system that also serves as a sort of skeleton; a digestive system to obtain, digest, and process food and to excrete waste; a nervous system to enable them to react to and function in their environment; and a reproductive system that provides more than adequately for the continuance of the race. They never rest, they do not have brains enough ever to be bored, and they can survive unfavorable periods either as eggs or by encysting.

The nematodes constitute a large and extremely varied group, commonly referred to as round worms (one authority estimates that the group contains 500,000 species) and they live in many different ways — as free-living nematodes in soil, manure, and decaying vegetable and animal matter, as parasites in plants, and as parasites in animals. One student of them says that probably there is not a single species of marine or land animal that does not at some time in its life harbor in its body one or more species of parasitic nematodes. However, these animal-parasitic nematodes are outside the scope of this chapter and will not be mentioned further.

As with animals, so with plants — there probably is no single species of higher plant that is not, at some time and place, invaded and injured to a greater or lesser extent by one or more kinds of nematodes. Most commonly the roots of plants are invaded, since most of these parasitic nematodes inhabit the soil, but stems, leaves, flowers, and fruits may also be invaded and damaged by nematodes. Some of these infestations may reduce crop yields almost to the vanishing point, others cause various degrees of injury or disfigurement. These plant-parasitic nematodes are especially prevalent in the tropics, as are so many other minute but vicious enemies of man, of his cultivated plants, and of his domestic animals, but no region where economic plants are grown is entirely free from damage by nematodes. A few examples will suffice to illustrate this.

There are probably more than a hundred major diseases of the cultivated potato, and one of these, caused by the potato root eelworm or golden nematode (so called from the gold-colored masses of cysts on infected roots), is of importance in Europe as far north as the British Isles and, in the United States, it has for some decades been established in potato fields on Long Island. It has been suggested that this nematode was brought to Europe on potatoes from South America when these were first carried to Europe by the Spaniards a few hundred years ago. However, it was not noticed in Europe until 1881, some 300 years after the potato was introduced and more than a hundred years after the potato had been widely grown and intensively cultivated in Europe. Potatoes were brought from South America to Europe not once, but many times, and the nematode might have been carried along many times too. But this nematode also infests some of the weed members of the potato family that are native to Europe, and it might have been living in the roots of these weeds long before the potato was introduced, just waiting for a more succulent and more valuable host.

Root and tuber crops, cereals, citrus fruits, pineapples, coconuts, bananas — all are subject to damaging attack by nematodes of one sort or another; 23 species of nematodes are listed by nematologists as attacking citrus fruits, 15 species are listed for banana plants, 36 species for tea plants, 23 species for coffee, 16 species for forest tree seedlings in nurseries, and so on. One nematologist says, "Vegetable production in virtually every part of the world is impaired to a greater or lesser extent by nematode pests." Nematodes even attack cultivated mushrooms and, if they do not attack wild mushrooms, they certainly are found on them, sometimes in tremendous numbers.

Although the study of nematodes parasitic in man and in other animals is in the domain of zoologists — either protozoologists or nematologists — the study of those which cause plant diseases is in the domain of plant pathologists or, more specifically, plant nematologists.

The fact that many serious diseases of economic crop plants were caused by nematodes was not discovered until the early 1930s, and then by plant pathologists. Since there were at that time no trained nematologists to take over the study of these problems, plant pathologists took on such study themselves, and in this way plant nematology became a branch of plant pathology. This is reasonable enough, since the damage done to plants by parasitic nematodes very often is increased and accentuated by the viruses introduced by the nematodes themselves and by the bacteria and fungi that accompany or follow the damage done by the nematodes, and to that extent some of the nematode-caused diseases of plants are really disease complexes, with a number of different kinds of organisms contributing to the total effects.

About the time that nematodes came to be recognized as important causes of disease in plants, and probably in part because of it, a mycologist in the United States Department of Agriculture, Charles Drechsler, became interested in some of the nematode-trapping fungi, and the study of these fungi became his major occupation and preoccupation for more than a decade. One of his summary papers is listed in the references for this chapter (72). Those who want an over-all introduction to the biology of these fungi are referred to the book on them by Duddington (73).

The fungi that trap nematodes do so basically by two methods: (1) snares and (2) adhesive pegs. Representatives of each will be described and illustrated in the pages that follow.

Nematode-Snaring Fungi

The snares produced by nematode-snaring fungi are of several different types. Most commonly they consist of a single circle or loop, or of a fairly complex latticework of interconnected loops. The same fungus may produce both single and compound snares. The circular snares may be constricting or nonconstricting; both types will be described shortly. One investigator of these fungi in the 1930s suggested that these snares must have arisen suddenly, by mutation, at some time in the past, arguing that they could hardly have evolved gradually, since an incomplete snare would be of no conceivable use. This is similar to the argument against Darwin's theory of gradual evolution through descent with modification, to the effect that something so complex as, say, a human eye could not possibly have evolved from some simple prototype because a primitive and poorly functioning eye could not have any survival value. Good logic, perhaps, but poor biology: many morphologically simple animals and plants, including the fungus *Pilobolus*, which is one of nature's marvels, have very simple light receptors — or eyes — that perform important functions in the lives of those that possess them, and so presumably have fairly high survival value. Considering the great diversity of design and function of the different kinds of snares, their sudden rise de novo would indeed be a biological miracle of the first order. We have no way of knowing what the precursors of these fungi and the precursors of the present-day nematodes looked like or how they lived a couple of hundred million years ago, but it seems highly probable that something so complex and so delicately adapted to their environment as are these nematode-snaring fungi must be the product of almost infinitely long and slow evolution.

The constricting snares consist of several cells, and when a nematode pokes his head or, more rarely, his tail, into one of these snares, the mechanical stimulus of contact causes the cells to swell instantly and to constrict the nematode in a death grip, as shown in figure 6-1. A nematode so snared will struggle and thresh around violently — he knows that he is in deep and desperate trouble. He occasionally will break the snare off from the parent mycelium and carry it away, but here again he is doomed, since a branch of the fungus soon grows out from the constricting snare where it is in contact with his body wall, penetrates into his insides, and there branches abundantly and digests him.

Whether the nematode enters these snares by chance or is attracted to them by some subtle scent is not known, but to one watching the process it seems almost as if the nematodes were baited into the snares by some means or other. A nematode will approach and lightly touch a snare with his snout, then recoil from it sharply, as if he had received an electric shock, retreat a bit, approach again, leave, return still again, and finally put his head in the noose. It just does not look like chance. This sort of trapping is not comparable to snaring rabbits or other animals that have a definite runway which they repeatedly use and in

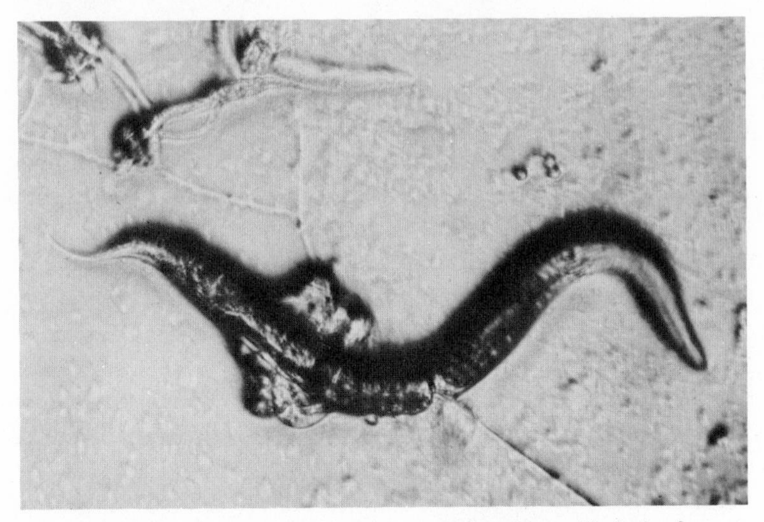

Figure 6-1. A nematode captured by a predacious fungus. The body of the nematode partly conceals the remains of a previously caught nematode, now almost completely digested by the fungus.

which the snare is placed. A better comparison would be catching fish in individual and unbaited snares fastened here and there on a fishline and let down into the lake. I have not tried that, but I am sure it would not catch many fish. The process as described above, and as described by many others, is observed on the surface of agar in a culture dish, not in soil. In the culture dish the nematodes are free to wander all over the surface of the agar, and they do so; if the agar is not too firm they will at times burrow down into it. In soil, of course, they are free to squirm about in three directions, and they do so, and in any soil where nematodes are moderately abundant, which is just about any soil containing decaying organic matter, there must be a thousand different kinds of microflora struggling for survival all the time. The nematode-snaring fungi are not present in every bit of soil or decaying plant debris that harbors nematodes in abundance, and even when these fungi are present they do not predominate by any means among the competing microflora; it seems highly unlikely that even under the most favorable conditions in soil there would be anywhere near the amount of mycelium of these nematode-trapping fungi present that we see on agar in petri dishes. All this argues against the idea that such snaring is accomplished by mere chance. On agar, even when the nematode population is not especially high, when the nematode-trapping fungus has only a few strands growing out from whatever material was put on one side of the dish, and when numerous other kinds of fungi are present, dozens of nematodes may be caught along one branch of mycelium in a few days. One sees the ''ghosts'' (the undigested outer integument) of the nematodes that already have been consumed, recently killed nematodes whose insides are filled with mycelium, those just caught and still struggling violently — all in no more than a one-fourth-inch length of mycelium. In some places there are literally bunches of trapped nematodes, in all stages from those freshly caught to those entirely consumed. There are just too many nematodes caught in too many snares in too small a space for it to be chance alone.

The nonconstricting, ring-type snare is relatively easily broken away from the hypha on which it is borne — it probably is so designed. The nematode pokes his slender head and the fore part of his body through the snare, and then in his struggles breaks the snare away from the parent mycelium. He may soon poke his head through another ring, and may also break this one loose from its moorings as he struggles. Duddington says he has seen a nematode bearing six such rings. These rings are barely large enough to span the girth of the nematode; in fact they fit so tightly that the nematode must, to some extent, have to force his way into them. Yet when a nematode is struggling to be rid of one such snare, he may poke his head into others. Again, it seems unlikely that this involves chance alone. Even a single snare, whether or not it is torn away from its parent hypha, means death to the nematode that bears it. As with the constricting snares, a branch soon penetrates through the body wall of the victim, ramifies throughout the body cavity, and within a short time digests the entire interior of the nematode.

There probably are still other kinds of nematode-snaring fungi, and other types of snares, than those that have been described. Several years ago, in a culture dish set up to get study material of this sort for my mycology class, a fungus appeared with snares that, when the nematode stuck his head into one of them, snapped into a figure eight and clamped the victim securely; I have not seen such a one described elsewhere.

If these nematode-snaring fungi are grown alone, in pure culture on ordinary agar, they do not form snares. The usual way to observe them in action is to pour sterile water agar (just water and agar, no nutrients) into petri dishes, allow it to harden, then put a bit of soil or decaying wood or leaf mold or manure on the surface of the agar at each side, and wait. Within a few days nematodes will be swarming over the surface of the agar, and within a week or so nematode-trapping fungi usually will appear. The nematodes never fail to appear in abundance, but sometimes, for reasons unknown, no nematode-snaring fungi grow out. The most reasonable explanation for this is that none happened to be present in the bit of material put on the agar.

At first these fungi will form only a few scattered snares, but once a nematode has been trapped, invaded, and consumed, snares are formed in abundance. By adding this or that compound to the agar medium in which the fungus is growing, snares can be induced to form in the absence of nematodes, but never in such abundance as when the nematodes themselves are being snared; obviously this whole business is biologically complex.

Some nematode-catching fungi do not make use of snares, but simply have sticky knobs on the end of short branches that grow out from the hyphae, or there may be just a sticky end of this short branch or peg — no knob. When the nematode comes in contact with this, the adhesive on the knob or on the end of the peg holds him so tenaciously that struggle, squirm, and thresh around as he will, he cannot get away. From the adhesive knob or peg a hypha grows through the body wall of the nematode, proliferates in his interior, and digests him.

This process seems straightforward enough on the face of it, yet there may be more involved than meets the eye, as so often is the case. The adhesive must be a peculiar one. The nematode, remember, is slick and slimy, and mostly muscle. He is swimming or wriggling in a thin film of water that always is present on the surface of the agar — a visible film, not a hypothetical one. Catching and holding a muscular, fighting worm under these circumstances by means of a small adhesive peg with a bit of adhesive on the end of it is roughly comparable, say, to catching a four-foot-long and four-inch-thick snake in the swamp with adhesive on the end of a pencil eraser, which is highly implausible.

Nontrapping Fungi Parasitic in Nematodes

The fungi described above are said to be predacious, because they catch their prey alive before they consume it; once the fungus has grown into the interior of the nematode its further growth is one of parasitism. A number of other fungi consume nematodes also, but without trapping or snaring or otherwise immobilizing them first. So far as is known, all the fungi of this latter type, regardless of their taxonomic position, are obligate parasites of nematodes, although there have been no serious attempts to grow them in pure culture. Also, so far as is known, these particular parasitic fungi parasitize nematodes only, and certainly none of them ever have been recorded in higher animals or man, which is a blessing. In general, the process is about as follows: A nematode foraging in soil or on an agar plate by chance consumes a few spores of one of these fungi; the spores pass into the midgut of the animal, germinate there, and, like the mycelium of the nematode-trapping fungi, digest the nematode from within. Soon after the nematode is dead, branches of the fungus grow out through the body wall and produce a small crop of spores, which in turn may be consumed by another nematode. Nematode-parasitizing fungi are found in all the several major taxonomic categories or groupings of the fungi, and from that standpoint this must be considered a fairly successful way of life. Whether anything more than chance is involved in the nematodes consuming spores of fungi that shortly will digest them is not known, but it is at least possible: spores of fungi are not one of the favored foods of nematodes; in fact although I have spent many hours watching nematodes and fungi in culture dishes, I cannot recall *ever* seeing a nematode consume a fungus spore — so perhaps the spores of these peculiar parasites have some fatal attraction for their victims.

Many of the plant-parasitic nematodes are exceedingly difficult to control, and naturally the possibility of using predacious fungi for this purpose has had a great appeal. As the trapping is observed under the microscope in the laboratory, it is dramatically final so far as the individual nematode is concerned; one moment the nematode is free and full of vigor, then suddenly he is in the toils of this flimsy but fatal fungus, choked and strangled into submission, and shortly consumed. Many of these nematode-snaring fungi can be grown readily on ordinary agar media or on autoclaved moist grain, where they produce tremendous numbers of spores, and so production of inoculum is no problem. It was thought that if such inoculum were incorporated into the soil of fields and gardens where damage by nematodes was occurring, the populations of nematodes might be reduced to the point where their depredations would be minor. No need for poisonous chemicals, no planting the infested fields with different crops for five to ten years, in the hope that this would reduce the population of nematodes harmful to the economically preferred crop, no expensive and not very effective fumigation — naturally such a prospect had an appeal.

As one reasonably direct approach to this end, various sorts of organic matter — manure, cabbage leaves, other plant remains — were at various times and places added to plots of soil, the idea being to furnish an environment favorable to the predacious fungi. Of course, the addition of those materials would stimulate the growth of many other kinds of fungi in the soil; whether or not it would increase the population of predacious fungi was an open question. The important thing was whether it increased the yield of the crop grown there, as compared with the yield of the same crop in soil not so treated. At times, some of these treatments have resulted in increased crop yields in relatively small test plots, but whether because of an increase in predation by fungus-trapping nematodes or for other reasons could not be determined. It is extremely difficult to manipulate an environment as biologically, physically, and chemically complex as the soil so that one organism, out of the thousands of kinds present, which happens at that particular time and place to be undesirable from the standpoint of growing a certain crop, will be selectively reduced or eliminated. That probably is one of the valuable biological principles that has been learned from such attempts at controlling nematodes. The experiments didn't work out as it was hoped that they would, but this could be determined only by actual trial.

Even under the artificial conditions maintained in a petri dish, which probably are much more favorable to the predatory fungi than conditions usually are in soil where crops are grown, the nematode population holds up very well. In a single culture dish, where a nematode-snaring fungus or nematode-snaring fungi can develop without much competition from other fungi, the nematodes are caught by the hundreds, but they multiply by the thousands, and the number of nematodes snared usually rises slowly to a peak and then slowly declines; the nematode population may be reduced below what it would be if no predacious fungi were present (even this is not certain, because there are other checks on population increase among the nematodes, too, and if the check in the form of predacious fungi were not operating, maybe these other checks would be operating) but there always are plenty of nematodes present. That in soil or compost or decaying plant and animal remains and manure both the hunter fungus and the hunted nematode survive is shown by the fact that, when we place bits of these materials on agar, nematodes soon appear, and they usually build up to a moderately dense population before the predatory fungus or fungi appear. One assumption made in some attempts at biological control is that if the population of injurious or damaging organisms — in this case parasitic nematodes — is reduced by 90 percent, the injury or damage to a crop is reduced by 90 percent. This is not necessarily true; it is, in fact, very likely to be grossly untrue: if there already are twenty times as many nematodes present as are necessary to cause a given amount of damage, then elimination of 90 percent of this population would still leave twice as many present as would be required to cause that amount of damage. Very seldom do any predators or parasites eliminate *all* of their hosts; if they did, they would eliminate their own source of food and thus would eliminate themselves. For this reason, biological control of pests of any kind is likely to be just partial control; very few undesirable plants or animals have been by this means reduced to the point of elimination. This general principle does not, however, seem to apply to desirable species; the valuable forest chestnut of the eastern United States has been virtually eliminated by the chestnut blight fungus, and in many areas of the United States east of the Rocky Mountains elms have been virtually eliminated by a combination of Dutch elm disease, caused by a fungus, and phloem necrosis, caused by a virus.

Fungi Parasitic in Insects

Insects, like other living things, are subject to infection by viruses, mycoplasmas, bacteria, and fungi, and some of the diseases caused by these parasites can, at certain times and places, greatly reduce the population of both injurious and beneficial insects. In general, the diseases of insects caused by fungi are much less destructive than some of those caused by viruses and bacteria. This is true of diseases of higher animals too — many fungi infect man and his domestic animals, but none of them decimate the population as did the epidemics of black death in Europe a few hundred years ago, or as did the smallpox virus among the plains Indians over a hundred years ago, or, more recently, as did the influenza epidemic after World War I. However, some of the fungus-caused diseases of insects are widespread and have interesting aspects; representative ones will be described.

The Species *Beauveria*

The fungus now known as *Beauveria bassiana* was first described in 1835, and the specific name *bassiana* was in honor of an Italian gentleman farmer and all-around scholar and investigator named Bassi, who found the fungus on silkworm larvae and who established the fact, by experimental inoculation, that the fungus infected and killed the silkworms. This really was a significant achievement, since, as always, a number of complications were involved, among them the fact that the fungus does not ordinarily sporulate on the infected larvae until after they die, and then only when the right relative humidity and temperature prevail. So infected and ailing but still living larvae were not infective, whereas dead ones often were. Also there were other diseases of the silkworm later shown by Pasteur to be caused by bacteria and viruses, and these must sometimes have confused the picture. Bassi began his work on this problem in 1807, when he and others thought that the disease was due to something nonbiological in the environment. By 1825 he had accumulated what to him was convincing evidence that the infection was due to a fungus, although many others refused to accept his conclusions and scoffed at his evidence. (Many people still scoff at evidence obtained through experiment, apparently preferring to believe that truth is established by revelation and logic.) In any case, he submitted a summary of his work to a commission of illustrious men, who reviewed it and put their stamp of approval on it, after which it was published, in 1835. It is available in English translation, as a volume in the Phytopathological Classics series (71), even though work on a disease of an insect is not strictly plant pathology. His work was the first to establish the fact that a fungus could cause a disease of anything, and in this he anticipated the plant pathologists who, about 1850, first proved that a fungus could cause a disease of plants. He also preceded medical researchers, including Pasteur, who had not yet developed the germ theory of disease, and so he deserves more credit than he is usually given for this work.

Bassi was not a mycologist and so he sent the fungus to a collaborator, who described it and named it *Botrytis bassianae*. Later other species in the genus, affecting other insects, were described, under many different names. Mostly they were called *Botrytis* in Europe and *Sporotrichum* in America, although, as a French mycologist showed in 1913, none of these fungi belonged in either of these two genera. In 1940 an American working in Cuba isolated a similar fungus from a dead root of a yucca plant (most probably from a dead insect in the root) and in spite of the by then extensive literature in several languages on *Beauveria*, he erected a new genus, *Tritirachium*, from the fancied resemblance of the spore-producing structure of the fungus to the zigzag rachis of the seed spike of wheat. Subsequently a Frenchman, who certainly must have been familiar with the earlier French publications on this fungus, and who had isolated still another species of it (this one from the infected eye of a girl in Cairo, Egypt), transferred *all* species of *Beauveria* to *Tritirachium*. What a crazy and pointless confusion! No wonder the working mycologist interested mainly in how fungi live and what they do sometimes views with despair the seemingly senseless antics of the nomenclaturists and taxonomists to whom the all-important thing is the name.

Beauveria bassiana has the distinction of having been the first fungus used in biological warfare against insects, although its use for that purpose was hardly intentional. An investigator of diseases of silkworms in France about 1850 had a brood, or part of a brood, of larvae infected with the fungus, and after the larvae died he threw a tray of them out the window. A tree just outside the window was infested with leaf-eating caterpillars. These became infected with spores from the fungus on the discarded silkworms, and within four days, so it is said, *all* the caterpillars in the tree were dead. That all of them died from this very casual and offhand method of inoculation is hard to believe, since no researcher has ever reported a 100 percent kill by artificial inoculation with this fungus; however, it makes a good story.

Different species of *Beauveria* infect hundreds of species of insects, and many of these insects are destructive to economic crops. At some times and places epidemics caused by this fungus will almost eliminate a local population of insects, and the idea of using *Beauveria* to produce epidemics among noxious insects occurred to many entomologists in the last century. Steinhaus (76) summarizes research on the biological control of insects by means of viruses, bacteria and fungi up to 1900, indicating that a good deal of work in this field had been done up to that time. *Beauveria* can be grown readily in pure culture and large quantities of spores are relatively easy to get; also they retain their viability for a fairly long time. A reasonable approach seemed to be to grow the fungus, harvest the spores, and spray or dust these in fields where crop plants were being eaten or otherwise damaged by insects known to be susceptible to infection by *Beauveria.*

This was done in the midwestern United States in the late 1890s and early 1900s — 50,000 packets of spores were distributed to farmers over a period of a few years in Kansas, the spores to be spread in wheat fields to control the destructive chinch bug. A similar approach was used in Russia in an attempt to control insects on sugar beets, and in other places to control still other insects. All in all, these constituted rather extensive tests.

No clear-cut evidence was adduced that insect mortality from this artificial inoculation of the fungus increased over that caused by the inoculum naturally present. If the conditions for development of the fungus were exceptionally favorable, including a fairly heavy population of susceptible insects (a heavy population would mean, of course, that considerable damage to the crop plants already had been done), infection by *Beauveria* might reduce the insect population drastically. Without the combination of circumstances necessary for an epidemic to develop, simply adding more inoculum had little or no effect.

According to the principles of epidemiology, establishment of an epidemic of any sort in any population of any kind of plant or animal requires (1) large numbers of susceptible host individuals, (2) inoculum of a virulent pathogen, (3) conditions for rapid development of the pathogen in the host, and (4) rapid increase and spread of inoculum. If any one of these is lacking it doesn't much matter how the others are manipulated — an epidemic will not develop.

After almost a century of work, much of it by some very sharp, knowledgeable, and imaginative men, no damaging insects anywhere are regularly controlled by artificial distribution of inoculum of *Beauveria*. Local epidemics that nearly eliminate a given species of insect in a small area are reported from time to time, but this fungus has not become a useful tool in mounting an effective campaign against our enemies in the insect world.

Fungi are not the only disease-causing agent tried against insects — viruses, bacteria, and parasitic insects have also been used, some with much more success than has attended the use of fungi. Other means have been found effective against certain insects: sterilization of the male screwflies by radioactive materials (the insects mate but do not reproduce); and chemical love potions (the chemical released by the female, which attracts the male from afar, is synthesized in the laboratory and spread around miscellaneously in the territory occupied by the insects, so the male flies around aimlessly and never finds the females, which soon pine away). All these have their limitations. The viruses, bacteria, and fungi and the parasitic insects and the insect-eating birds constitute the main natural checks that keep any one species of insect from taking over the world, but each of these has its own natural checks as well, so that in the resulting complex interplay there sometimes is an outbreak of army worms or locusts that destroy our fields and forests, and there sometimes is an outbreak of one or another disease that greatly reduces the population of insects. It seems inevitable that insecticides will continue to form a major means of control of many kinds of noxious insects for a long time to come.

Fungus Diseases of the Honeybee

The lot of the honeybee is not always a happy one; it is subject to dysentery, constipation, and paralysis, and it even can be afflicted with intestinal stones. The hives are invaded by wax moths that destroy the combs, and the bees within the hives are parasitized and plagued by lice, mites, flies, and ants, and in their work outside the hive they are attacked by spiders, wasps, hornets, and birds. One of these enemies is a wasp called Bienenwolf (= beewolf in German), the female of which descends upon the busy bee in flight, paralyzes it with an injection of poison, carries it home (the beewolf's home), and stacks it in the larder for her young to feed on when they hatch. (There may be friendly coinhabitants in the hive as well as unfriendly ones; one of these is a miniature "scorpion" — actually a small spider with its front legs modified into great lobsterlike claws with which it snatches up moth larvae, lice, and other evil intruders and eats them.) Bees also are susceptible to attack by *Beauveria* and by another fungus parasite, *Ascosphaera apis*, but the damage done by these two fungi is not really very great. The Swiss keep records of this sort of thing, as they evidently do of so many other sorts of things, and over a period of 19 years in Switzerland, including the decade of the 1920s, only 64 cases of infection of hives by *Beauveria* were reported by beekeepers, whereas there were 1377 reported cases of American foulbrood, a bacterial disease.

Ascosphaera apis is not known to occur in the United States, but is found in Europe and in the British Isles. It presumably occurs naturally on pollen, and is carried into the hive on it, along with a multitude of other fungi. It may infect and kill some of the larvae but does not menace the survival of the colony as a whole, and it usually dies out and disappears without treatment. Bees must be rather difficult to treat in any case, and until very recently much of the treatment must have been on a hit-and-miss basis. A British writer about 1920 listed the following "medicines" as having been fed to healthy colonies to keep them healthy, and to diseased ones to cure them: beef tea, carbolic acid, formalin, naphthol, "Izal" (a proprietary compound containing heaven only knows what), alcohol (this does the beekeeper good, and so presumably should help the bees, too), onions, Bulgarian sour milk, vinegar, salt, and jalap. It certainly does not sound as if the Age of Reason had by that time descended heavily enough upon the beekeepers to weigh them down unduly.

Beauveria and Ants

Some of the many kinds of ants that inhabit and bid fair to inherit the earth are either noxious or obnoxious to us and probably to quite a few other fellow travelers, and some of them appeared to investigators to be ideal candidates for control by means of the fungus *Beauveria*. Their nests were found to be humid, the individuals numerous, crowded together, and in constant contact, and the temperatures in the nests high enough to favor rapid development of the fungus. The only thing lacking for the development of a raging epidemic, seemingly, was an abundance of inoculum of the fungus, a lack that could be very easily supplied. The fungus was grown, its spores were harvested and added in quantity to ant colonies, but no epidemic developed in any of the experiments carried out. Why not? Some ants were infected by the original load of inoculum, and died, but as soon as they died their carcasses were carried out of the nest and disposed of, and so there was no subsequent increase of inoculum within the colony.

Fungi Parasitic on Scale Insects

There are many kinds of scale insects, so called because the young, soon after they hatch, settle down and secrete a tough chitinous cover or scale over themselves for protection from predators and from rough weather. Each individual scale insect sticks its proboscis down into the sap-conducting vessels of the twig or branch on which it has settled and sucks up juices for nourishment, often to the great detriment of the health and vigor of the host. These scale insects particularly relish some of our most valuable fruit trees and ornamental shrubs and, because of the protective shield beneath which they live, are difficult or impossible to control by means of even rather potent insecticides. In the tropics and subtropics where parasitic insects constitute an important natural control of scale insects, application of an insecticide may greatly reduce these enemies of the scale insects and allow the scale insects to increase wonderfully.

A number of fungi are parasitic on scale insects, and again the possibility seemed good that some of these fungi, if increased and sprayed on the scale insects in orchards, might greatly reduce or eliminate the damage caused by the scales. This was tried in a few orchards in different parts of the country, and artificial increase of the fungus enemies of the scale insects did result in some control, at least for a time, but in general it has not been successful. Petch (74), who for many years worked with fungi parasitic in insects, in 1920 said:

. . . at the present day after 30 years' trial there is no instance of the successful control of any insect by means of fungus parasites. If the entomogenous fungi already exist in a given area, practically no artificial method of increasing their efficacy is possible. If they are not present, good may result from their introduction, if local conditions are favourable to their growth, but, on the other hand, their absence would appear to indicate unfavourable conditions. . . . The problem which has yet to be solved by those who wish to control insects by means of fungi is how to create an epidemic when an epidemic would not occur naturally. The evidence indicates that it is not possible to accomplish that by the mere introduction of the fungus or by spraying spores from natural or artificial sources.

When the Biological Warfare Service was established in the United States after World War II it was regarded by some people with horror — a perversion of humanity, something far more inhumane than the agencies concerned with the traditional methods of killing and maiming. The idea of using biological warfare against fellow humans is, of course, repugnant, but in practice the activities of this organization seem to have turned out to be relatively humane. Most of the many plant pathologists in the Biological Warfare Service worked on some of the basic aspects of epidemiology of rust diseases of cereal plants, which doesn't sound especially horrid. In any case, our biological war against insects, waged with vigor and against a foe totally unable to fight back, has not really accomplished very much.

The Genus *Cordyceps*

The genus *Cordyceps* has more than 100 species, nearly all of them parasitic in insects of one sort or another — flies, wasps, ants, scale insects, spiders, moths, and assorted bugs and beetles. (See figure 6-2.) A few grow in underground fruit bodies of another fungus, an *Ascomycete*, the mycelium of which grows in association with the absorbing roots of conifer trees in our forests. This is a rather peculiar host range — insects and the underground fruit bodies of another fungus. About the only thing the two diverse host groups have in common is that some of the insect larvae parasitized by *Cordyceps*, as well as the fruit bodies of the fungus host, are underground. Maybe that is enough.

The ordinarily visible portion of *Cordyceps* is an upright stalk, tech-

Figure 6-2. Fruiting stalks of *Cordyceps* growing up from the remains of a parasitized insect larva

nically called a stroma, or a group of such stalks, growing from an insect. Representative examples are shown in figures 6-3, 6-4, and 6-5. Beneath the surface of the upper portion of each stalk are hundreds or thousands of perithecia, the fruit bodies in which the asci are borne. The asci in turn bear ascospores, and when these are mature they are forcibly discharged. The ascospores are long and threadlike, and just before they are shot out each spore divides into many short sections, each of which supposedly functions as a separate spore. What happens to them then we do not know, but they must germinate and establish mycelium in the soil or in one or another stage of their insect host. In any case, the mycelium eventually invades, kills, and digests the insect host and then sends up its spore-producing stroma or fruiting stalk. Some species of *Cordyceps* have an imperfect or conidial or asexual stage, usually in the genus *Isaria*, also characterized by upright stalks on which the spores are produced.

Many species of *Cordyceps* are tropical, but a few are common as far north as northern Minnesota. At the time the clump shown in figure 6-2

was found, similar clumps appeared, all of them on just about the same day, at an average of about one clump per square yard, over an area of about half a square mile in a mixed hardwood forest at Itasca Park in northwestern Minnesota. All of them that were dug up were growing from June beetle larvae. In that case the fungus must have been generally present in the soil over a fairly large area.

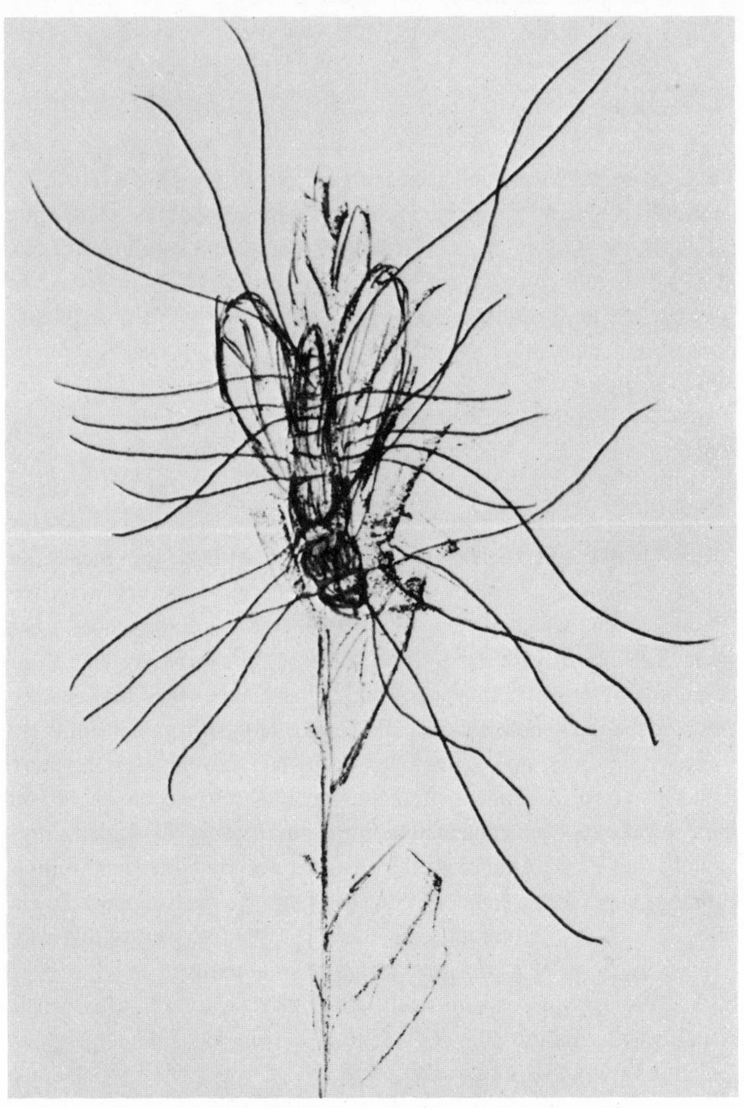

Figures 6-3, 6-4, and 6-5. Fruiting structures of *Cordyceps*-like fungi growing out from parasitized insects

The combination of an insect larva — or rather the mummy of a larva converted into fungus mycelium — and the fungus stalk was known centuries ago, long before the nature of fungi was known. One such combination was known as ''Chinese plant worm'' (it has since been named *Cordyceps sinensis*); this was supposed to have potent medicinal qualities. Specimens were rare, occurring only in the far reaches of

Tibet, where they were collected and sent to the Imperial household, to be used by the royal physician in the treatment of royal ills. One recipe called for the stuffing of a duck before cooking with specimens of this plant worm; during cooking the medicinal virtues of these were said to pass into the meat of the duck. Up until at least the 1950s this Chinese plant worm could be bought in herb stores in the Chinatown section of San Francisco.

In 1749 Torrubia, a Franciscan friar in Cuba, described and illustrated (or his collaborator illustrated for him) "trees" growing out of the bellies of wasps — sometimes the genus *Cordyceps* is known as *Torrubia*, in his honor. Some of these "trees" were said to be five "hands" high, a hand presumably being then, as now, four inches, so that the plant was 20 inches tall. Another account says they were five "spans" high, which would be about 30 inches. Either those were giant-sized wasps with giant-sized stalks of *Cordyceps* growing from them, or, more likely, things got out of hand somewhere between the field and the final published account. Copies of the illustrations of these "trees" still exist, and neither the wasps nor the fungus stalks look especially large. Specimens of an Australian *Cordyceps* are said to be more than a foot tall, and maybe they are. It is mildly aggravating, though, that all these fabulous things either occurred anywhere from a couple of hundred to a couple of thousand years ago or else appear only in some far-off land three-fourths of the way around the world and way back in the bush, and probably only once in a lifetime.

One tropical wasp is said to fly about and conduct its usual small business with a fungus club of *Cordyceps* growing out from the back of the joint between the head and thorax. By the time this fruiting stalk grows out in that way, almost all the insides of the wasp have been digested and converted into *Cordyceps*; the insect must be relatively insensitive to this sort of thing.

The *Hirsutella* illustrated in Figure 6-5 occurs on various species of wasps in the tropics; when first described it was thought to be a part of the wasp from which it grew, an idea which even a crude microscopic examination would have shown to be erroneous. Notice that the insect in death has come to rest near the upper end of the twig, which of course is advantageous to aerial dissemination of the spores of *Cordyceps*.

That is, the parasitized insect does not just keel over anywhere at random, but fastens itself, just before death, in a location preferred by the insensible fungus. Crickets parasitized by a species of *Empusa*, in the group to be taken up next, normally live and die hidden away beneath rocks and logs, but as they are about to expire from infection by this fungus they climb up on grass stems in the open and get as high above the ground as they can. How much psychology these insects are endowed with is questionable, but in any case both of these insects are induced, by some chemical produced by the fungus within their bodies, to alter their normal behavior pattern in a way that is beneficial to the dissemination of spores of the fungus.

Figure 6-6 shows a parasitized scale insect with fruiting structures of *Cordyceps* growing from it.

The Genera *Empusa*, *Entomophthora*, and *Massospora*

All three of these genera are closely related to one another, and all of them parasitize many kinds of insects. They have been observed and studied mainly in houseflies and their relatives, and in locusts, because in these insects they sometimes appear in epidemic form, killing the hosts by the hundreds or by hundreds of thousands. The three genera are separated from one another mainly on the basis of spore shape and how the spores are borne — the manner of living and the life history are essentially the same in all: a spore lands on a susceptible insect, germinates, penetrates through the body wall of the insect and into the body cavity; mycelium develops mightily there, digests and consumes the interior of the host, and converts this into another crop of spores. Resting spores —zygospores — may also be produced within the dead insects, and these spores are capable of enduring winter, drouth, and long periods of dormancy.

The big infection in houseflies and in some of their relatives occurs mainly in the fall. The infected insect may appear normal up to shortly before death, when it begins to move slowly and erratically, then soon comes to rest, often on grass blades or on walls or windowpanes, and it may be anchored securely there by a weft of mycelium that grows out of its underside. The fungus, which by then has consumed the entire insides of the insect, sends hyphal stalks out through the body wall,

Figure 6-6. Fruiting structures of *Cordyceps* growing from a parasitized scale insect

between the abdominal segments. A spherical or oval spore is formed on the tip of each stalk and is shot off. The spore is sticky, and adheres to whatever it lands on; the halo around the fly in plate 5 is made up of a mass of spores that were discharged within 24 hours after the fly expired, during which time several or many crops of spores must have been produced. Obviously a considerable number of spores had to be discharged to make up a deposit such as that shown, and probably even a greater number were shot off into the air, but these spores are rather large as fungus spores go, and so the deposit or the total crop of spores from a given fly probably would be in the hundreds of thousands, not hundreds of millions or hundreds of billions, as in some of the more prolific fungi. If a spore lands on another fly it supposedly germinates, causes infection, and soon kills its host and produces another crop of spores, but whether it actually is that simple may be questioned. This aimless and random discharge of spores off into the blue would seem a poor and chancy way of perpetuating the disease and the species — the chance of another fly cruising by within range at just the right moment must be almost astronomically remote. A spore of *Empusa* that lands on something other than a possible host has another opportunity — if the air is humid it can germinate and produce another stalk on which another terminal spore is formed and shot off. Even so, the likelihood of this one encountering another host insect in its brief trajectory of an inch or less must also be very remote. Yet we sometimes encounter, on the lawn in a humid fall, a group of 40 or 50 flies, in an area a foot or so across, all with masses of grayish spore stalks growing out from and covering most of the abdomen. All these flies must have become infected at about the same time. And also, since these particular flies are not normally especially gregarious, they must have become infected in different places. Only when about to die from infection by *Empusa* do they sometimes congregate, perhaps for mutual solace and commiseration, but more likely because this serves some biologically useful if obscure need of the fungus. How this fungus lives through the winter in northern climes is not known, but one possibility is by means of resting spores formed within the body of the dead insects. Dead insects where? Are these flies that congregated on the grass in the lawn when they were dying of *Empusa* infection going to stay there through the rains of fall and the snow and ice of winter, and all through the next summer, and indeed, if the weather next year is not favorable for the fungus to spread, through another annual cycle? With all the minor kinds of life constantly foraging for food on and near the surface of the ground, and a multitude of sharp-eyed birds on the lookout for the provender, plus the ravages of weathering, it seems highly unlikely that the fragile, fungus-filled insect remains would last through the winter. We just don't know what goes on. It would be interesting to find out.

The survival of one species of *Entomophthora* is even less explainable, if it is reasonable to have comparative degrees of explainability. This case is as follows: In the northern United States there are intermittent outbreaks of the forest tent caterpillar on forest and shade trees. In the aspen forests that cover much cutover land in northern Minnesota such outbreaks are likely to occur every seven to ten years or so, sometimes over a considerable geographic area at the same time. One of the enemies of these voracious caterpillars is a fly that larvaposits — that is, it deposits minute larvae, not eggs — on the caterpillars. These larvae chew their way into the insides of the caterpillars and consume them, and if these flies are present in sufficient numbers that may be the end of that particular infestation of forest tent caterpillars. The few that survive eventually fall to the ground where they change into pupae, which also have some enemies, and finally the next spring the surviving pupae change into the adult stage and these emerge as moths, mate, and lay eggs that hatch into larvae. The adults also have enemies and biological checks, and so it may require seven to ten years or so for the population to increase again to the point where it constitutes a damaging outbreak.

Why don't these parasitic flies continually keep the forest tent caterpillars in check? The flies in turn have enemies, one of which is a fungus, a species of *Entomophthora*. One year in the late 1940s there was an outbreak of the forest tent caterpillar over a fairly large area in northern Minnesota, from Itasca Park on the west to beyond Duluth on the east, a distance of more than 200 miles, and in a band about 100 miles wide. In late June the nearly full-grown forest tent caterpillar larvae were consuming aspen leaves voraciously, and they were so numerous that one with keen ears could hear the delicate, incessant patter of their dry pellets of excrement falling like miniature hail. Suddenly the parasitic flies appeared in abundance, and by June 28 they were so numerous as to constitute a nuisance — they larvaposited not only on the forest tent caterpillar larvae, their proper host, but also on the laundry on the line and even on the hands and faces of professors.

By July 3, by which time the larvae of the forest tent caterpillars were dying like flies, the parasitic flies also were dying like flies, and scarcely a living one was to be seen. But the twigs of hazel brush and other shrubs were thick with dead adult parasitic flies from which a fungus was growing. Some of these by now mummified flies were put on a glass slide in a moist chamber in the laboratory, and within 24 hours the fungus had shot off masses of spores. The same phenomenon — infection and death of the adult parasitic flies — prevailed throughout the entire area of approximately 20,000 square miles that embraced the outbreak of forest tent caterpillars, and it occurred within just about the same time span, which means that all of these uncounted billions of flies had to have been infected by the fungus at about the same time. The flies probably were about spent in any case, since their biological mission had been accomplished in their furious three-to-five-day bout of larvapositing, but where had the fungus been lurking during the seven lean years, and where did all the inoculum come from to infect simultaneously so many billions of flies throughout so large an area? Again, it would be interesting to know.

Massospora and the Seventeen-Year Locust

The wasp mentioned above as being able to go about its business with its insides converted into *Cordyceps*, and with a fruiting club of the fungus growing out the back of its neck, is outdone by a locust. The periodical or 17-year locust, or cicada, is just that: a brood of adults appear every 17 years; in their brief three months' appearance above ground they feed, mate, and produce a crop of eggs that soon hatch into larvae; these feed for a time, then drop to the ground, burrow into the earth, transform into pupae, and, sixteen years and nine months later, emerge as adults to repeat the cycle.

In some years when a given brood emerges, the males, soon after they have mated and so have fulfilled their role in the scheme of things, and soon to die anyway, become heavily invaded by *Massospora*. Some females may become infected by the fungus too, but by far the greatest number of obviously and visibly infected individuals are males. Mycelium of the fungus develops within their bodies and digests their entrails. The fungus produces masses of spores within the abdomen of the infected insects, and this causes successive segments of the abdomen to slough off, permitting the powdery spores to be disseminated as the insect moves about. Even when nearly all of the abdominal segments have been shed and there is only a hollow cavity from the last remaining segment right up to and into the head of the locusts, they still go about their regular business, feeding and probably chatting among themselves about the weather and laughing and making silly remarks about those hard-working ants storing up food for winter. No one knows how and where the fungus spends the intervening 16 years and nine months. The weather may not always be favorable for the fungus to produce a lot of inoculum every time that brood appears, so maybe it has to endure for 32 years and six months, or even longer.

The Genus *Laboulbeniales*

The *Laboulbeniales* include more than 1500 species, all of them obligate parasites of insects — they grow only on living insects. Actually only a relatively few kinds of the approximately one million described species of insects (there are probably two million or more total species of insects) have been examined for the presence of these peculiar fungi, so there may be more than the 1500 species so far described.

The life history of these fungi is relatively simple. They produce only ascospores. If an ascospore lands on or is depositied on a suitable site of a susceptible insect, it germinates, an infection peg penetrates through the host cuticle, and, so far as infection is concerned, that is it. Some species form a few short rhizoids, but most do not — there is only an infection bulb below the cuticle of the insect and no mycelium within the body of the host insect, or anywhere else for that matter. The insect parasitized in this way suffers no detectable injury, no impairment of bodily function, no decrease in life-span, and probably very little if any inconvenience.

In all of the *Laboulbeniales* there are two sexes, either on separate plants arising from separate spores that carry genes for maleness and femaleness respectively, or from separate branches of a single plant or single thallus. The male plant, or the male portion of the bisexual plant, produces single-celled, nonmotile spermatia, and these are carried to a receptive cell, a trichogyne (litterly, a "female thread") on the female plant or on the female portion of the bisexual thallus. After fertilization, perithecia are produced by the female plant and within these perithecia asci and ascospores develop. The ascospores when mature either are shot out or ooze out or are forced out by external pressure, as by one of the limbs of the male when he grasps the female during copulation.

The distribution of these peculiar fungi seems to be without rhyme or reason. A couple of years ago one of the students in mycology at the University of Minnesota Biology Station in Itasca Park undertook, as his project, the description of *Laboulbeniales* found there. In five weeks of at least moderately intensive hunting he did not find a single specimen. Yet some years before, in a colony of oriental cockroaches maintained for experimental purposes in the Entomology Department at the University of Minnesota, infection by *Laboulbeniales* on the antennae of the cockroaches was so heavy that the antennae looked fuzzy, and several species of the fungus were present. The infection in this colony disappeared as suddenly as it came, and did not subsequently return.

Some species of *Laboulbeniales* infect only one species of host, some infect several species of host and an individual insect may be infected with several species of *Laboulbeniales*. Some species of the fungus occur on just about any and all parts of the host, and some have a very restricted distribution on the host. One species, for example, is found almost without exception ". . . only on the humeral angle of the right elytron of the females and only on the inner distal surface of the femur of the right front leg of males" (75). This is specialization with a vengeance. The ascospores of this species are expelled from the perithecia only when pressure is applied from without — in this case by the leg of the male clasping the female during mating. For the infection to be passed to the next generation there must be some overlapping of generations and some promiscuity.

Ecologically, so far as anyone can see, these *Laboulbeniales* are of absolutely no significance in any way to anything but themselves, a situation that, in biology, must be almost unique. It seems strange that so large a number of species should have taken up this way of life, but judging from the diversity of species involved both as hosts and as parasites the relationship must be ancient, and biologically successful.

7

Fungi Pathogenic in Man and Animals

A few fungi long ago became adapted to a life of parasitism on the outside of or within the bodies of higher animals, including man. These fungi are not especially numerous in the way of species, but some of them have a wide host range and a wide geographic range, and are common in the sense of being regularly present. The diseases they cause are not usually fatal but can be irritating and disfiguring, and in one region or another at one time or another they may constitute a public health problem of some importance. These fungi and the diseases they cause are likely to be with us for a long time to come, and if the accounts of some of them border on the revolting, that happens to be the way it is — there is no way that this sort of thing can be made to appear pleasant or amusing.

In general, these diseases of warm-blooded animals caused by fungi are known as mycoses. The diseases themselves are classified in various ways. One of these ways, as reasonable as any other perhaps, divides them into (1) those involving the superficial covering of the body — the skin, nails, and hair; (2) those involving the deeper tissues, often referred to as deep-seated mycoses. These latter may involve the skin also but are not limited to the skin. Examples of each will be discussed.

Superficial Mycoses or Ringworms

One author says that ringworms probably are of great antiquity, a statement with which probably few biologists would quarrel, since if we and the fungi and all the innumerable other fellow travelers evolved together over the past three billion years or so, the various relationships among us, including parasitism, must have become established long ago. In any case, ringworms were common enough and conspicuous enough to have been included in compilations of medical literature made two thousand or more years ago. There is some evidence to indicate that ringworm diseases may have been much more common in the moderately recent past than they are now. In the times of the Tudor rulers of Great Britain, for example, some of the courtiers and hangers-on were given official permission to remain covered in the royal presence because they had ringworm of the scalp (and probably pretty badly, too) and looked less repugnant with their heads covered than they did with bare heads. At one time in the nineteenth century special schools were maintained in Paris, France, for children with ringworm of the scalp; presumably this was to segregate them after a fashion in the hope that they would not communicate the infection to their fellow pupils who were or appeared to be free of it; that the people in charge went to the trouble of setting up special schools for those so afflicted suggests that ringworm was prevalent, and probably also that it was more than just a transitory problem, here today and gone tomorrow.

Ringworm infections at times were prevalent in places other than the crowded and unsanitary cities such as Paris undoubtedly was when the ringworm schools were established. William Dampier was an English adventurer, freebooter, and pirate when England ruled the seas (he commanded the ship from which Alexander Selkirk was put ashore on the island of San Fernández — Selkirk's adventures there were later to be immortalized by Defoe in *Robinson Crusoe*; later Dampier himself was put ashore from another ship, and on another island, in the Indian Ocean). In the late 1600s, on an exploring and plundering voyage around the world he stopped in Mindanao, a village in the southern Philippines, and, among other things, described in his log a skin affliction that troubled many people there. It was a dry scurf that developed over portions of the arms, legs, and trunk of the victims, sometimes in circular and concentric zones and patches of roughened, scaly, and discolored skin, the scales or patches at times overlapping in such a way as to suggest a shingled roof, as shown in figure 7-1.

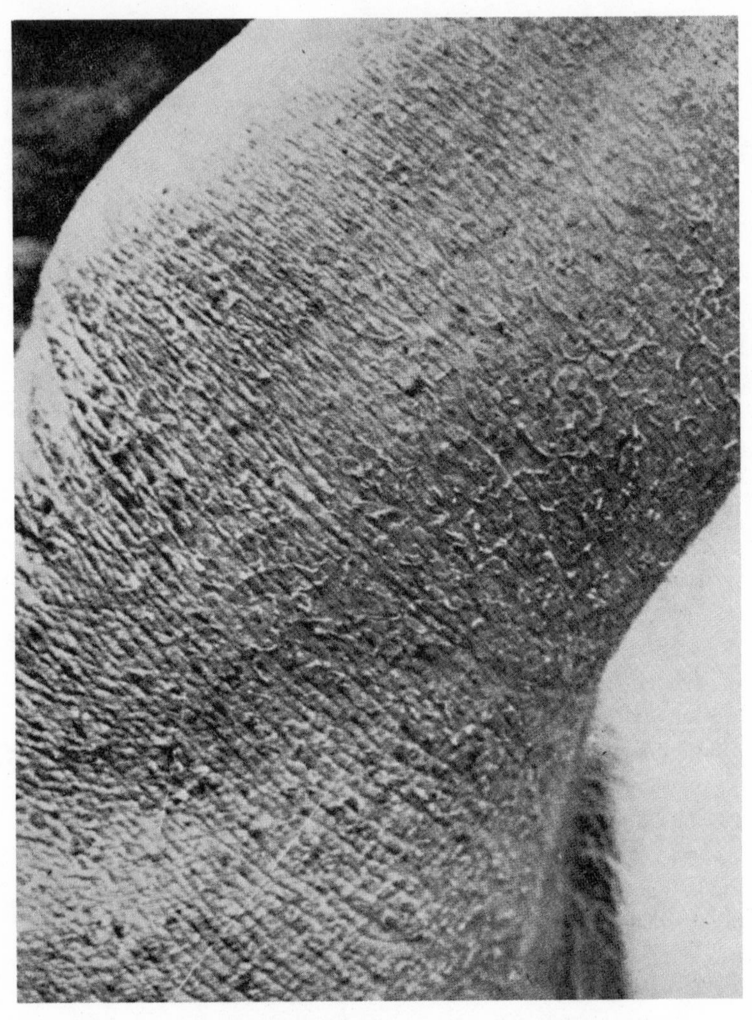

Figure 7-1. A ringworm-caused concentric scaling lesion on a patient's arm (courtesy of Dr. V. Pardo-Castello; reprinted from Conant *et al.*, reference 80)

Dampier said that it ''causeth great itching,'' which probably was an understatement; what he did not say was that, once this infection is established in the skin, it is likely to persist for years or even for as long as the person lives. The disease evidently was fairly common in the region then, and it evidently still is common there, not only in Mindanao, but throughout much of India, Southeast Asia, and Polynesia; indicative of its wide distribution are the names applied to it — Tokelau, Lafa Tokelau, Burmese Ringworm, Chinese Ringworm, Indian Ringworm, Scaly Ringworm, Tinea Circinata Tropical, and others.

According to Ainsworth (77), in 1879 a British medical officer stationed in Amoy, China, a coastal city of then minor commercial importance, proved, by inoculating one of his Chinese assistants, that this horrid disease was caused by a fungus. He may have seen fungus mycelium in microscopic mounts of shed scales of skin from a person or persons afflicted with this ringworm, and may have inoculated some of these scales into or onto the skin of his assistant, and so produced the disease, but this is hardly "proof" that the disease was caused by a fungus. There may have been and probably were a dozen different organisms in such skin scales, since most of us carry a moderate array of microflora even on healthy skin, and dozens of kinds can be isolated from moribund skin; all he proved was that the disease was transmissible. To establish that a given fungus or bacterium is the cause of a given disease requires a good deal more work than just transfer of infected tissue from a diseased to a healthy individual, followed by development of typical symptoms.

In the 1945 book by Conant *et al.* (80) this particular ringworm is said not only to cause intense itching, as Dampier mentioned in the late 1600s, but also to be very "resistant to treatment," a euphemism meaning that once you get it there isn't much that can be done to rid you of the infection — either you get well spontaneously or you have to live with it. Under the circumstances, one wonders how the medical assistant inoculated by his superior in Amoy felt about the matter, and whether he suffered "great itching" over most of his body for the rest of his life. If so, it must have been a heavy price for him to have paid for this somewhat dubious research. At least the medical officer evidently did not inoculate himself, which showed reasonable foresight and caution on his part — maybe he was saving himself for other great advances in the study of communicable diseases in the future. He could have inoculated some kind of laboratory animal other than his Chinese assistant, but, as it happens, this particular ringworm fungus, originally known as *Trichophyton concentricum*, but now known as *Trichophyton schoenleinii*, is pretty strictly adapted to the human skin and rarely can become established in the skin of laboratory animals, even after inoculation onto or into the skin, and so inoculation of laboratory animals would have yielded no information whatever concerning the transmissibility of the disease.

The fact that ringworm in man was caused by fungi was at least fairly well established, and some of the fungi responsible were even named, in the 1840s, by medical men in Europe; they evidently had plenty of good material to work with. Gruby, for example, a Hungarian working in Paris, named one of these common ringworm fungi *Trichophyton*, meaning "thread plant," a name that remarkably enough it still retains. This was some 30 years or more before techniques were developed to grow fungi and bacteria in pure culture on sterile agar media: Gruby must have been a fairly sharp person — and probably also lucky, since he had no proof that the fungus he named and described actually was the cause of the ringworm with which he found it associated. Some of the ringworm fungi form rather characteristic mycelium and spores on infected skin and hair, and this must have been what Gruby saw and described as the cause of the infection in the skin. The actual proof that ringworms were caused by fungi came decades later when work with pure cultures got under way.

Ainsworth, already cited, says that ". . . more than 350 species of dermatophytes have been proposed and given approximately 1,000 names," a somewhat confusing statement. The actual number of names does not matter of course, and this is mentioned only to show the reader that these fungi have been given a multitude of different names, which means that many different men have worked with them in different times and places. Naturally, many of the workers who described this or that ringworm fungus were familiar only with the particular isolate of the particular species of fungus that they happened to have, and probably not too familiar even with that isolate. They grew it in culture, studied it briefly, and published a research paper on it. (Researchers traditionally have been and to a large extent still are judged, by their administrative superiors, on the *number* of papers they publish, not on the quality of the contents; those researchers who almost never publish deplore this, and those who publish frequently applaud it, as do most of those who supply funds for research. After all, if a researcher has been working hard for ten years and has not yet accumulated enough new information to justify the publication of even a brief paper, then what has he done to advance knowledge?) Each newly isolated ringworm fungus was given a name, sometimes according to the symptoms with which it was associated in that particular patient, or according to the part of the body affected — the top of the head, the neck, the face, the hand, the arm, the leg, the foot — or according to the geographic region where it was encountered, or sometimes according to whim. The thousand names that got into the literature by this haphazard approach were not the product of thorough and enlightened mycological study, and many of these supposed contributions to medical and mycological knowledge contributed very little to knowledge, but only to confusion.

Georg (84) at the Communicable Disease Center in Georgia, where probably more and better work has been done on these fungi and on some aspects of the diseases they cause than has been done anywhere else in the world, lists four species of *Microsporum* and eight species of *Trichophyton* as constituting the total number of dermatophytes or skin-inhabiting fungi, plus an oddball *Keratinomyces ajelloi*, isolated only once, from a squirrel in a Chicago zoo. Workers at the Communicable Disease Center have the advantage of having received and studied isolates of these fungi from all over the world, and so they have been able to determine the range in genetic and environmental variation among them, which is essential if one is to make any real sense in taxonomy, although many taxonomists still are happily unaware of this basic principle — or, if they are aware of it, they certainly disregard it. As one view on the vital importance of this interminable name-changing that goes on in mycology Ainsworth, speaking specifically of the ringworm fungi, says that to freeze the nomenclature of these fungi would be "intolerable" to the systematists, who must retain their self-appointed and inviolate right to devise new systems and coin new names. In other words, the devising of new systems of classification and the coining of new names for these by now well-known and relatively stable fungi are to continue into the foreseeable and probably into the unforeseeable future — any other alternative would be "intolerable" to the systematists. Many of these systematists and nomenclaturists obviously are not overburdened with common sense. They profess to be seeking the final ''natural'' system, that holy grail of systematists. So far, it has eluded them. Simpson, in his *Life of the Past* (86), emphasizes the need for recognition, by classificationists, of evolutionary and biologically realistic concepts and procedures, but says that some students still do not grasp this, and some are unaware of it. He also emphasizes that what the classificationist is dealing with and classifying is *populations* of organisms, not just individual specimens. It may seem somewhat strange that these words of taxonomic wisdom should come from a paleontologist who deals with the fossil remains of plants and animals long since dead, yet it may not be strange at all; his everyday work and his whole professional career enabled him to see the grand sweep and course of organic evolution through populations changing with time. The total biological experience of some of those who coined many of the 1000 specific names for these few dermatophytes was limited to the single isolate of the fungus that they chanced to have. If retention of a classification of these fungi into 12 species, in two genera, is ''intolerable'' to the systematists, it certainly is very highly tolerable to the great majority of those who are concerned in one way or another with what these fungi do and how they live.

In spite of the fact that several ringworm diseases were described in the 1840s and were properly attributed to infection by fungi, they were very little studied until the 1940s, when ringworm and other fungus infections in military personnel in the humid tropics of the South Pacific called expert attention to the problems. One of the products of this was the book by Conant *et al.* already referred to; since that time, a good deal of attention has been devoted to these diseases, in human medicine and in veterinary medicine.

Epidemiology of Ringworms

Ringworm can and does occur almost everywhere geographically, but the cases of really horrid skin infections that disfigure and digest the features from the outside in are relatively rare outside the tropics, and are not too common there; they do constitute another ever-present hazard in the unfriendly humid tropics — where life, according to historian Arnold Toynbee, who never tried to subsist next to nature there, is so easy that people are indolent and backward from lack of the need to struggle. In many areas of the tropics the diseases of man caused by viruses, bacteria, fungi, protozoa, worms, and other kinds of parasites are so prevalent, so draining of all energy, physical and mental, that mere survival must often be the end-all of existence. No one suffering from a combination of the many ills endemic there, as well as being ill-fed on a diet deficient in proteins and vitamins, and being verminous and dirty, is going to devote much time to Higher Things. Many of the tropical nations are underdeveloped precisely because the conditions that prevail there make development extremely difficult. In some tropical areas man can be no more than a temporary, unwelcome, and unsuccessful intruder.

The fungi that cause ringworm have been divided into three groups, based chiefly on their host preferences, as follows:

1. Zoophilic or "animal loving." These infect animals primarily — dogs, cats, horses, cows, poultry — but are rather readily transmissible to man. The fungi in this group are *Microsporum canis*, *M. distortum*, *Trichophyton verrucosum*, *T. mentagrophytes*, *T. equinum*, and *T. gallinae*.

2. Anthropophilic or "man loving." These infect man primarily and cannot readily be transferred to animals. This group contains *Microsporum audouinii*, *Trichophyton tonsurans*, *T. rubrum*, *T. violaceum*, and *T. schoenleinii*.

3. Geophilic or "earth loving." The single species in this group, *Microsporum gypseum*, occurs naturally in soil, presumably as a saprophyte; it can infect animals and man.

Zoophilic ringworm fungi. Dawson (81) discusses ringworm problems in cattle, horses, pigs, sheep, goats, fowls, dogs, cats, chinchillas, foxes, hares, mink, muskrats, nutria, squirrels, mice, rats (many investigators use and handle and live fairly close to white rats in the course of their work on nutrition, mycotoxicology, and so on, and may get ringworm from them), guinea pigs, rabbits, hedgehogs, porcupines, and primates. The list probably could be extended to just about anything with hide, hair, or feathers. I have seen no reports of superficial fungus infections of frogs, toads, turtles, lizards, or snakes, but given the environment in which many of these animals live such infections must occur at least occasionally. Workers at the Communicable Disease Center in Atlanta, Georgia, obtained samples from more than 2000 individual animals throughout the United States, and plated these on agar to determine if ringworm fungi were present; fungi capable of causing ringworm were found in more than one-fourth of the samples — 27.2 percent, to be exact. They also obtained hair specimens from 2350 wild animals caught at random and similarly plated the samples to detect the fungi present; fungi known to cause ringworm were isolated from 3.2 percent of these specimens. None of these wild animals from which ringworm fungi were recovered had any skin lesions or hair abnormalities — they were, in other words, symptomless carriers. Some domestic animals are symptomless carriers of some of these fungi also, and so, most probably, are some people.

The paper by Georg quotes veterinarians to the effect that one cat in five and one dog in ten are infected with ringworm. These fungus diseases in domestic and zoo animals are of sufficient importance in veterinary medicine to have, at least in some colleges of veterinary medicine, a special course or courses devoted to them. A book by Jungerman and Schwartzman (85), *Veterinary Medical Mycology*, is also tangible recognition of the importance of these diseases in animals. Many veterinarians, by the way, in the treatment of a pet with ringworm, do not want the animal in their hospital or clinic were it can spread the infection among their other patients, but prefer that it be kept in the owner's home.

Infection is spread mainly by spores, but also presumably by mycelium in skin scales and in fragments of hair. Some of these ringworm fungi grow and form masses of spores on the outside of and within the hairs on the affected part, as well as on and within the skin. These spores are very durable and can remain alive for years in blankets, in contaminated straw or bedding, and in brushes, combs, and other grooming tools. People who regularly work with or handle animals almost inevitably must be rather frequently exposed to infection by these fungi, and many infections of ringworm in the owners of pets have been traced directly to infection in the pets. These ringworm fungi are not killers, but when an epidemic develops in a given flock or herd or stable it may be more than just unsightly — the animals may be made so uncomfortable with constant irritation that their productivity is reduced.

Some of these fungus infections can be detected by the use of a so-called black light — under ultraviolet of a certain wavelength the infected hair or skin fluoresces brightly. Some, however, cannot be detected in this way, the fungus does not fluoresce, and positive diagnosis requires microscopic examination of the hairs and skin to detect the characteristic spore masses and plating fragments of hair and skin on appropriate agar media to isolate pure cultures of the fungus or fungi. The standard agar medium for this purpose was for many years Sabouraud's agar, named after the Frenchman who developed it, shortly before 1900. This has since been greatly improved upon by the addition of antibiotics that inhibit the development of bacteria likely to be present in hair or skin infected with fungi.

These diagnostic tools and techniques for the detection of infections by ringworm fungi probably are used much more routinely in veterinary diagnostic laboratories and in small animal hospitals than they are in hospitals and clinics dealing with human diseases, in part perhaps because doctors of veterinary medicine are more alert than are doctors of human medicine to the possibility of fungi being involved, and in part to the fact that use of the facilities of diagnostic laboratories manned by expert diagnosticians and pathologists has come to be an integral part of the practice of veterinary medicine.

Anthropophilic ringworms. According to Henrici (87) ringworm, especially of the scalp, at one time was a very common disease, particularly in children of the poorer classes. The inference usually is that this was mainly a matter of such children being exposed to less soap and water than were the children of the well-to-do. Very likely they were, but that this made them more susceptible to ringworm is no more than a guess, and probably a poor one. They undoubtedly also had a diet deficient in many things that the body needs to ward off different kinds of infections. Certainly there never has been any suggestion that susceptibility or resistance to ringworm in animals and epidemic outbreaks of ringworm in flocks or herds are related to the lack of skin cleanliness in the animals concerned.

Ainsworth says that ringworm is considered by the middle- and upper-class mothers in England to be a somewhat disreputable disease (so evidently the children of at least some of these mothers must get ringworm) and mentions that large outbreaks of ringworm caused by *Microsporum* may originate when the children in school handle the school cat. Whether the school cat actually was shown to be the source of infection or was only conjectured to be is not stated. The following is a warning concerning ringworm issued by the Department of Education in St. Paul, Minnesota, in 1945, indicating that ringworm of the scalp was present in epidemic form among schoolchildren in many cities of the United States.

RINGWORM OF THE SCALP

An infectious scalp disease is spreading among school children in many American cities. This epidemic type of ringworm is comparatively new in this community and may cause permanent baldness. Children are chiefly affected, especially those under sixteen years of age.

The outstanding signs are the presence of patches of partial baldness, stubby hairs due to breakage, lack of lustre of infected hairs, and varying degrees of scaling and inflammation.

Unless it is recognized early and treated properly, certain types of ringworm of the scalp may spread relatively rapidly and even assume epidemic proportions. In order to reduce the spread of the disease to a minimum, it must be recognized *early*.

Most infections are probably spread at play, although the theaters and barber shops contribute to the extension of the epidemic. Infection is also spread by articles of clothing, especially caps and scarves, and by combs, brushes, towels or anything that comes in contact with the infected areas.

Schools are being surveyed with the Wood Lamp and cases found are excluded from school.

Parents are urged to observe the following precautions in trying to prevent the spread of the disease:

1. Keep your children away from known cases.
2. Watch scalp carefully.
3. Children should wear a washable head covering if attending a movie and should be instructed not to rest their heads on the back of the seat.
4. Shampoo head thoroughly with soap and water after every visit to a barber shop.

5. There should be no exchange of combs, brushes or articles of clothing between children.
6. Inspect family pets — cats and dogs — for infection and consult a veterinarian if suspicious symptoms are found.

Should your child become infected, take him promptly to your family physician. He must have a permit from the Bureau of Health before returning to school.

He should not be allowed to attend movies, Sunday School, church or any public gathering, and must not mingle with other children.

He should wear a head covering that can be boiled or burned, and sleep alone to avoid infecting other members of the family. Known cases should wear a head covering at all times, and old caps and hats that cannot be washed and boiled should be destroyed. Scrupulous cleanliness in the home is of great importance.

George W. Snyder, M.D.
Director of School Hygiene

Certainly this epidemic, or these epidemics, of ringworm that developed more or less simultaneously in many cities of the United States were in no way associated with school cats, and, judging from the warning given, the infection was known or believed to be highly contagious from person to person. Also there was no hint that it was associated with lack of personal cleanliness. The reasons for these occasional and sometimes widespread flareups of ringworm of the scalp among school-age children are not known. Once the infection is established in a child it may persist for years, but it usually disappears spontaneously when he reaches the age of puberty.

Several species of fungi that cause ringworm are common among adults, and it seems highly probable that some of them are present in the skin of this or that portion of the body more or less regularly without causing any obvious symptoms — the fungus or fungi that cause athlete's foot, for example, must be about as widely distributed among the population, including nonathletes, as are the viruses that cause the common cold. The same fungus or fungi sometimes infect the nails of the feet and hands and literally rots them — a most offensive affliction, and one that does not readily yield to treatment. The groin, chest, and bearded areas of the face are other favorite sites of infection by these ringworm fungi. There must be great differences among individuals in susceptibility to infection by these ringworm fungi, and a great difference in the susceptibility of an individual from time to time. There probably are differences in pathogenicity among strains or races of the different species of the fungi too, since this is the rule in biology, especially with such widely distributed organisms as these ringworm fungi. Presumably infection is spread mainly by spores, since some of the fungus species produce large numbers of spores, and from the nature of the spores produced — dry and powdery — it seems likely that they are airborne (which is why veterinarians do not want ringworm-infected animals to remain in their clinics or hospitals, and why the warning from the St. Paul Department of Education, just quoted, recommends that children with scalp ringworm should wear a head covering at all times). If inoculum is airborne all of us must at some time or another be exposed to infection by various ringworm fungi, and those who work with or associate with dogs and cats must be regularly exposed to infection — yet relatively few become infected. Why does infection develop on one portion of the scalp and not on another? Once the infection is established, there must be enough inoculum to cover the whole scalp and the rest of the body every day. Sometimes one person in a family, or one animal in a herd, will get ringworm and it will not spread to others, whereas at other times it is highly contagious. There is still a good deal to be learned about ringworm.

Geophilic ringworm fungus. The single fungus, *Microsporum gypseum*, in this group has been isolated at least fairly commonly from some soils, and the presumption is that it is present in many more places than where it has been found — there has been no really intensive search for it in soils by men competent to isolate and to recognize it if it were present. There are so many fungi present in soil that unless a technique is available that will disclose the presence of a given one, or of a given group, it is impossible to say that that one, or that group, is *not* present. And of course if one cannot find it, he cannot say that it *is* present. Whether *M. gypseum* is rare or prevalent in soil, it is a relatively uncommon cause of ringworm in either man or domestic animals, with the exception of dogs, in which it is said to be common.

Systemic or Deep-Seated Mycoses

A dozen or more species of fungi cause various systemic or deep-seated infections in man and animals, and several of these, chosen because of their prevalence or because of some of the interesting aspects of the fungi or of the diseases, will be described.

Coccidioides Immitis and Coccidioidomycosis

The fungus *Coccidioides immitis* causes primarily a respiratory disease in animals and man, but from the focus of infection in the lungs it may spread throughout the body by way of the bloodstream and cause pathologic changes — lesions of one sort or another — in just about all tissues in all parts of the body. In the usual course of events, infection results in a more or less acute but benign and self-limiting respiratory disease, and once the patient recovers from this he is likely to be permanently immune from further infection. Fiese (83), an authority on this disease, says that about 60 percent of those infected have few or no symptoms, and 40 percent have symptoms of varying degrees of severity — chills, fever, chest pains, coughing, lassitude — symptoms typical of a dozen other infections as well. These symptoms develop ten to fourteen days after infection, and may persist for some time, but eventually, in most cases, the immunological processes of the body take over and rid it of infection, although lesions and scars may remain in the lungs. In a relatively small percentage of the cases — one in 500 or so — the fungus "disseminates" from the lungs to other parts of the body, and this secondary stage may result in increasingly severe lesions in the skin, bones, and internal organs, so that the victim becomes a mass of external and internal lesions and abscesses. Once this stage has been reached, recovery is not likely; death may occur within weeks or after a long and lingering illness. Sometimes the disease proceeds to a fairly advanced stage and then remains static for years, and it may regress and then later reappear. Figure 7-2 shows an advanced case.

Whether the relatively innocuous and usually minor primary infection will progress to the exceedingly noxious and usually fatal secondary stage depends in part upon the race of those infected, as shown in table 7-1, from Fiese. The death rate of blacks from infection by this fungus

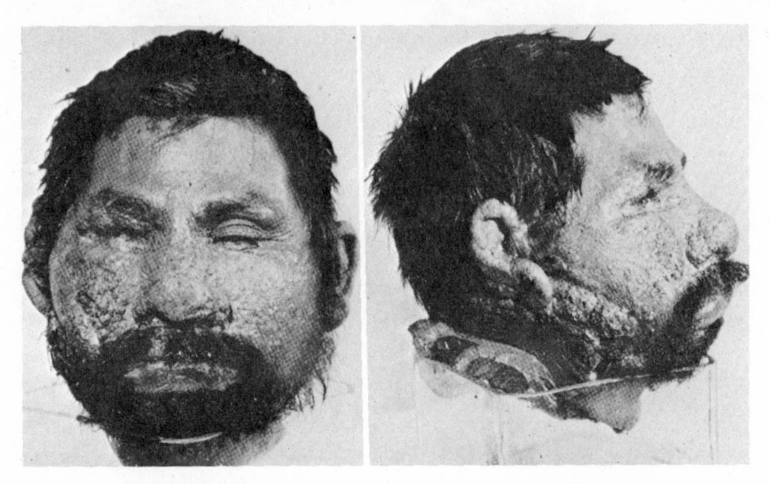

Figure 7-2. A man infected by *Coccidioides immitis*; his was the first described fatal case (reprinted from Fiese, reference 83)

was more than 23 times that of Caucasians, and the death rate of Filipinos was nearly 200 times as high as that of the whites. These figures were gathered in Kern County in the San Joaquin Valley of California, where the disease is endemic and where, judging by skin tests, more than 90 percent of the longtime residents have at some time been infected by the fungus. Men are more prone than women to develop disseminated infection.

The first case of coccidioidomycosis was described in Argentina shortly before 1890; the patient (shown in figure 7-2) suffered for seven years before he died. The pathologist on the case attributed the infection

Table 7-1. Comparative Morbidity and Mortality Rates for Disseminated Coccidioidomycosis among Various Racial Groups in Kern County, California, 1901–1936

Race	Case Rate per 100,000	Case Rate Ratio	Death Rate per 100,000	Death Rate Ratio
Caucasian	82	1	28	1
Mexican	281	3.4	140	5
Black	1,122	13.7	654	23.4
Filipino	14,350	175	5,381	192

Source: Adapted from data of Gifford, Buss, and Douds as recorded in Fiese (83).

to a protozoan parasite resembling *Coccidia*, whence the name *Coccidioides* — Coccidia-like. He inoculated dogs and monkeys with infected tissue and produced the disease but did not isolate the fungus on artificial media. The first case in the United States was reported in 1894 in California. The investigators plated out pieces of diseased tissue, but when a fungus grew from them they thought it was a contaminant and discarded the cultures. They were not the first, or the last, to throw away valuable results because the results were not in agreement with a preconceived notion of what ought to be. A short time later other investigators established the fungus *Coccidioides immitis* as the cause of the disease — they isolated it in pure culture, inoculated it into laboratory animals, produced typical symptoms, and then reisolated the fungus from the diseased animals.

That *Coccidioides immitis*, when first observed in diseased tissues, was thought to be a protozoan is easily explained; in such tissue it does not usually form mycelium but grows by means of peculiar sporangialike "spherules." In culture on agar it grows by means of mycelium, and aerial branches of the mycelium fragment into small, one-celled powdery spores that are easily airborne. Those who work with cultures of this fungus should handle them with respect. Fiese, in fact, says that "*Handling* COCCIDIOIDES IMMITIS *without proper precautions is foolhardy — and for members of certain races may be suicidal*." This seems a bit strong, but he should know. Some of those who worked with cultures of this fungus before its infectivity was known, and who described its characteristics in culture, evidently did not use any special precautions — at least they do not mention any; perhaps they were just lucky. Fiese cites a case where several medical students working in a ward two floors above the laboratory where cultures of the fungus were being handled became infected, presumably from that source. That is not so implausible as it may seem, since as described in chapter 5, spores liberated on the first floor of a building were carried through the three floors above in large numbers within minutes. Fiese also cites other cases in which laboratory personnel developed serious infection from careless handling of cultures of *Coccidioides immitis*. Certainly any fungus that can do to the human body what this fungus can do must be handled with care. One wonders how the doctors who first isolated the fungus and who considered it to be only a contaminant disposed of the cultures.

In the regions where the disease is endemic the fungus is found commonly in the soil, and especially in the soil in and near rodent burrows, and in the rodents themselves. At one time it was thought that these rodents furnished the main reservoir of inoculum, but now it is thought that they are just chance hosts, as is man, of what is primarily a soil-inhabiting fungus — and a fungus that probably was of little or no biological significance at all until men who happened to be susceptible to it moved into its territory in large numbers.

Its territory is the borders of deserts — some deserts, not all deserts. It may survive in many out-of-the-way places (although it is not known to) but to thrive it requires a desert climate, very hot and dry through part of the year but with a period of fairly heavy rain in the winter. The long, hot, dry season eliminates nearly all fungi from the upper layers of the soil, *Coccidioides immitis* among them, but these fungi survive in the soil eight to twelve inches below the surface. The wet season permits *C. immitis* to grow and sporulate in the upper layers of the soil before its competitors can become established. It can grow, in this fashion, in sandy desert soil, where there is little competition from other fungi, but it does not thrive in soil containing much organic matter, where other fungi are numerous. There may be other and more subtle ecological adaptations too, since there usually are, but if so they are unknown. In areas where the fungus is endemic, infections in man are much more numerous soon after the end of the wet season than at other times. Since the fungus is, in nature, a soil inhabitant, agricultural workers in the endemic regions are more exposed to infection than are those of other occupations. There is some circumstantial evidence that infection may result from spores on cotton, wool, and fruits and vegetables in general (some people handling these things regularly have become infected) but, so far as anyone is aware, fruits and vegetables shipped out of the San Joaquin Valley never have been implicated in the spread of infection of *C. immitis*.

Up to about 1910 nearly all cases of coccidioidomycosis were encountered in California, in part because it was common there and in part because some of the physicians there recognized it and could diagnose it correctly. For a time it was referred to as the "California Disease" and as "San Joaquin Valley Fever," which brought harsh criticism from the editor of the Bakersfield *Californian* (Bakersfield is in the valley) who referred to this as "infamous advertising." In the time from 1951 to 1955 nearly 400 patients were hospitalized in Kern General Hospital in Bakersfield with at least moderately severe to severe coccidioidomycosis, so there must have been quite a few more with lesser infections by the same fungus. However, the disease is by no means limited to the San Joaquin Valley, or to California — the endemic area in the United States extends eastward through Arizona, New Mexico, and the western half of Texas. Fiese says that there must be five or ten thousand primary infections per year in Arizona, with a very high incidence around Phoenix (see tables 7-2 and 7-3). In World War II

Table 7-2. Incidence of Disseminated Coccidioidomycosis in California, 1903–1954

Year	Cases	Deaths	Year	Cases	Deaths	Year	Cases	Deaths
1893.....	1	0	1914.....	5	4	1935.....	28	14
1894.....	1	1	1915.....	2	1	1936.....	44	17
1895.....	0	1	1916.....	5	3	1937.....	38	27
1896.....	0	0	1917.....	4	3	1938.....	71	19
1897.....	0	0	1918.....	1	1	1939.....	73	21
1898.....	0	0	1919.....	11	4	1940.....	54	25
1899.....	2	1	1920.....	5	3	1941.....	41	17
1900.....	2	2	1921.....	8	7	1942.....	49	27
1901.....	3	1	1922.....	4	3	1943.....	31	23
1902.....	0	1	1923.....	16	4	1944.....	40	17
1903.....	1	0	1924.....	17	8	1945.....	43	21
1904.....	1	0	1925.....	8	4	1946.....	43	20
1905.....	2	1	1926.....	14	6	1947.....	63	20
1906.....	3	2	1927.....	19	7	1948.....	72	52
1907.....	3	1	1928.....	36	11	1949.....	91	60
1908.....	1	0	1929.....	46	25	1950.....	92	66
1909.....	0	2	1930.....	22	11	1951.....	64	43
1910.....	1	1	1931.....	19	8	1952.....	64	43
1911.....	4	1	1932.....	19	9	1953.....	91	56
1912.....	3	5	1933.....	54	25	1954.....	78	37
1913.....	4	3	1934.....	49	29			

Source: Adapted from Fiese (83). Data for the years before 1928, when coccidioidal granuloma was first made reportable, are from Special Bulletin No. 57, California State Department of Public Health, and the dates are those of diagnosis, not of onset. Data from 1942 onward include both civilian and military figures. Compiled by Ida May Stevens, chief morbidity statistician, California State Department of Public Health. Courtesy of Dr. C. E. Smith, president, California State Board of Public Health.

Table 7-3. Reported Incidence of Coccidioidomycosis in Arizona, 1942–1956

Year	Cases	Deaths
1942	57	0
1943	222	0
1944	43	0
1945	7	1
1946	17	2
1947	5	6
1948	15	1
1949	89	2
1950	180	2
1951	180	7
1952	386	5
1953	97	8
1954	207	8
1955	164	7
1956	573	6

Source: Adapted from Fiese (83). Compiled from data supplied through the courtesy of Clarence G. Salsbury, M.D., commissioner, Arizona State Department of Health.

military training camps were established in the San Joaquin Valley and in areas of Arizona, New Mexico, and Texas where the disease is present, and many soldiers developed coccidioidomycosis of various degrees of severity. At one air force base near Phoenix, it was said by the medical officers that more man-days were lost from coccidioidomycosis than from the other three most frequent illnesses combined plus those lost from disabling injuries. The endemic area extends across the border into Mexico, and there are endemic areas in Venezuela, Argentina, and Paraguay in South America.

In the army camps, with constantly changing personnel being exposed to infection as new batches of men came in for training, the incidence of the disease supposedly was greatly reduced by oiling or paving the camp grounds, including the playing fields. Presumably the men in those camps would still inhale the dust blown into the camp from outside, so if the major source of infection is soil-borne spores, these must be relatively short-lived once they become airborne or the dust suppression within the limits of the camps would not have reduced the incidence of infection as it apparently did.

There is no effective treatment of coccidioidomycosis in man or in animals.

Histoplasma Capsulatum and Histoplasmosis

The fungus *Histoplasma capsulatum* causes histoplasmosis in man and in dogs, rarely (or at least rarely diagnosed) in other domestic or wild animals. The first three cases of histoplasmosis were described in the Panama Canal Zone in 1905 and 1906 by a young medical pathologist, S. T. Darling. The patients had died of massive infections, and in postmortem examination of diseased tissues he found this strange and evidently new organism that he thought was a protozoan, and to which he gave the name *Histoplasma capsulatum*. The first case in the United States was recorded in 1926, in Minnesota, in an area where infection by this fungus in the human population subsequently was found to be very common. Up to 1934 only six cases had been described, in Panama and in the United States, all postmortem. In that year Dr. W. A. DeMonbreun at Vanderbilt University isolated the fungus from an infected patient, grew the fungus on agar, and described its characteristic spores. He also inoculated the fungus into laboratory animals and recovered it from them before and after they died.

Up until that time and for nearly another decade histoplasmosis was thought to be a rare and almost invariably fatal disease, and relatively little attention was paid to it. In the early 1940s many men who were given chest X rays as part of their medical examination to determine their fitness for military service were found to have calcified pulmonary lesions indicative of healed-over infections, the incidence of these lesions being especially high in men from the Mississippi and Ohio River valleys; few of these men reacted to tuberculin tests and so it was unlikely that the lesions were due to tuberculosis infections. By then some knowledge of *Histoplasma* was being accumulated, and a histoplasmin skin test had been developed to aid in detecting those who had been infected by the fungus — a test similar to the tuberculin test for tuberculosis and to the coccidioidin test for coccidioidomycosis. From 1945 on, fairly widespread and large-scale use of this skin test revealed the fact, almost totally unsuspected by most investigators up to that time, that a large number of people in some areas of the United States reacted positively, although they appeared to be perfectly healthy; at some time in the past, whether they knew it or not, they had been infected by *Histoplasma capsulatum*. The book *Histoplasmosis* (88) states that it is likely that as many as 20 percent of the population of the United States are or have been infected by the fungus, and that there may be more than half a million new cases per year. The great majority of these either have no symptoms at all or suffer only miscellaneous aches and pains, with a light cough, perhaps some dysentery, very much like the symptoms of coccidioidomycosis, flu, and various bacterial infections. These symptoms soon disappear, and the individual is then highly resistant to or totally immune from further infection by the fungus. In most cases this high degree of resistance or immunity is life-long.

In a very small percentage of cases the fungus spreads, by way of the bloodstream, from the original focus of infection in the lungs throughout the body, and this may result in massive and fulminating infection that usually is rapidly fatal. The *Manual* by Conant *et al.*, already cited, says that up to then (1945) fatal cases of histoplasmosis rarely were diagnosed before autopsy. They probably were rarely diagnosed after autopsy either, since up to the early 1940s there probably were few pathologists with the training and experience to recognize the fungus in sections of diseased tissues, and still fewer who could isolate and grow the fungus on agar and recognize it. Medical mycology was then still in its relative infancy.

So far as the survival of the patient was concerned, the lack of accurate diagnosis probably made relatively little difference, since no effective treatment for histoplasmosis was available. Once the patient developed the disseminated and progressive form of the disease there was not much that could be done for him. There still isn't.

The fungus. DeMonbreun, who first isolated *Histoplasma capsulatum* from a patient and grew it on agar, suggested that the fungus might be found growing saprophytically in nature (82). On what he based this suggestion is not known, but since the fungus was not at all fastidious in its growth requirements, as some pathogenic fungi are, and could be cultivated on a variety of ordinary agar media at room temperature, the suggestion was a reasonable one. And, as it turned out, not only reasonable, but right.

The first to isolate *Histoplasma capsulatum* from soil was Dr. C. W. Emmons, a mycologist with the U.S. Public Health Service. Epidemiological studies in some communities in northern Virginia showed that 83 percent of the people reacted positively to histoplasmin skin tests, and 41 percent had calcified pulmonary lesions, findings which indicated a high incidence of infection in the human population in the area. Also *H. capsulatum* had been isolated from a small percentage of more than 2000 wild Norway rats that had been trapped in the area, and also from seven other species of animals; in addition nearly 50 percent of the healthy dogs and cats tested there had benign histoplasmosis. Since the disease is not communicable from man to man, from animal to animal, or from animal to man, it seemed highly probable that a reservoir of inoculum of the fungus existed rather generally in the area. One likely place to look for it was in soil, since soil harbors such a great variety of fungi. Moreover, *Coccidioides immitis* already had been shown to have its source of inoculum in soil, a fact with which Dr. Emmons was very familiar since he had isolated that fungus from soil in and near rodent burrows and from rodents themselves in the California desert.

Emmons devised a very selective isolation technique to detect *Histoplasma capsulatum* in soil: He suspended small samples of soil in sterile water to which antibiotics had been added to suppress the growth of bacteria. A small portion of the suspension was then injected into mice, which are susceptible to infection by this fungus. After four weeks the mice were sacrificed and tissues from their liver and spleen were plated on an agar medium and incubated at 30° C (86° F). If *H. capsulatum* was present in these tissues it would grow out of some of the samples. Since *Coccidioides immitis*, as mentioned above, already had been found in the soil of rodent burrows, one reasonable place to look for *H. capsulatum* also was in soil in and near rodent burrows. After more than 150 samples had been put through the isolation technique described above without recovering the fungus, it finally was recovered, from soil near a rat hole at the side of a chicken house. The significant thing, as it turned out, was the chicken house, not the rat hole. The fungus was found to inhabit, as its favorite ecological niche, soil containing the droppings of birds and bats. The fungus grows there, and presumably sporulates there, although, so far as I know, spores have not actually been detected in soil; possibly it survives in soil mainly as mycelium, and possibly mycelium fragments constitute the main form of inoculum. In any case, infection results from inhaling dust containing spores or mycelium or both.

The soil in and near old, dirt-floored chicken houses is likely to harbor the fungus, as is the soil under roosts where pigeons or other kinds of birds congregate, and soil in bat caves. The fungus was found in soil near the foundation of a house where bats roosted under the loose siding, but was not found in soil a few feet away. Bats themselves may harbor the fungus and carry it in their migrations.

Once the fungus has developed in a given soil it may persist and remain infective for years, so that if the soil is covered, then later again exposed it still can cause infection. Epidemics of histoplasmosis among dwellers in a given apartment house complex, and among pupils of a given school, have been traced to such bird-dropping-contaminated soil that was covered when the buildings were constructed and, some years later, again exposed. As would be expected, infections are more common in rural areas, where people live closer to the soil than do those in cities. Many examples could be cited, but perhaps one will suffice: several individuals in a farm family became infected with histoplasmosis when they cleaned out an unused silo in which pigeons had roosted. The implication is that they had not previously been exposed to infection, or at least not to a heavy enough one to have resulted in their developing immunity, even though they lived near plentiful inoculum. That the fungus is present and even fairly common on many farms is indicated by the fact that city children who were negative to histoplasmin skin tests before they visited farms reacted positively shortly after they returned from brief visits to farms.

Epidemiology. There are unexplained peculiarities in the distribution of histoplasmosis. In one area, 85 to 90 percent of the people may react positively to histoplasmin skin tests and, one or two hundred miles away, only 5 percent of the people may react positively. The disease is endemic in the Mississippi and Ohio River valleys, but there are local and epidemic areas far removed from that general region, and there are occasional local outbreaks here and there. The obvious explanation of this somewhat erratic pattern is that inoculum of the fungus is abundant in one area at one time and scarce there most of the time, and that the fungus may produce a heavy load of inoculum in one area at one time but does so only sporadically. Inoculum of *Histoplasma capsulatum* has been caught from the air but it seems unlikely that the spores would be present in the air in large enough numbers to cause epidemics more than a very short distance from where they originate. No one knows how many spores are required to cause infection in man, although it is said that inoculation of mice with ten spores regularly resulted in infection. In most of the outbreaks in a given family, or in a given small group, that have been traced to their source, infection resulted from exposure for some time to a rather high concentration of infective dust. The disease appears to be worldwide in distribution, but it is much more common in some areas of the United States, Mexico, and South America than it is elsewhere, and in some countries it never has been recorded.

Diagnosis. Positive and unequivocal diagnosis of infection by *Histoplasma capsulatum* involves histoplasmin skin tests, the more complex and time-consuming serological tests, identification of the fungus in sputum, blood, or diseased tissues, and isolation and identification of pure cultures of the fungus from the same sources. None of these procedures are exactly straightforward and simple. In the histoplasmin skin test, for example, if the weal resulting from injection of the histoplasmin is less than five millimeters (about 1/5th inch) in diameter it is rated negative; if larger, it is rated positive. This arbitrary size is based on experience and judgment, of course, but it is not infallible. Some people who have or who have had histoplasmosis do not react positively to histoplasmin, and some people who have been infected with other fungi, but not with *H. capsulatum*, do react positively. Recognition of the fungus in sections of diseased tissue is not always easy. It will not be present in every cell in diseased tissue by any means, and the diseased tissue may be invaded by a variety of other organisms. Even when it is present some expertise is required to find and recognize it. The fungus cannot always be isolated easily from diseased tissue either — sometimes many samples must be plated to obtain a single culture, so if the fungus is not recovered from diseased tissue in such platings the investigator does not know whether this is because it was not present or simply because it did not grow out. What this all adds up to is that histoplasmosis is not always easy to diagnose, either in the relatively mild and almost symptomless cases or in the more serious ones, and positive diagnosis requires more facilities, experience, and expertise than are likely to be available outside of well-staffed laboratories.

Candida Albicans and Candidiasis

Candida is a yeastlike fungus, growing by means of filamentous mycelium from which yeastlike cells bud off and which then continue to grow by budding, like the "true" yeasts. The quotation marks around the "true" are to indicate that no one really knows what a true yeast is, although a lot of people think they do. Henrici (87) says, "Many bacteriologists with little experience in studying yeasts think that they know very precisely what a yeast is and define it as a unicellular fungus multiplying by budding. Actually such a definition will apply to only a small proportion of the organisms usually classified as yeasts and only to these when they are maintained under constant conditions and not studied too closely." There are yeasts and yeasts, and they cannot all be fitted or forced into a sharply delimited pigeonhole. The group now known as *Candida* once was known as *Monilia*, and the disease now known as candidiasis once was known as moniliasis. Ainsworth (78) says there are 30 species of *Candida*. Some of these are widespread in soil and on the surface of fruits and berries and seeds — those are the natural sites of the yeasts used to convert the juices of these fruits and berries into fermented wines and beers.

Most cases of parasitism by *Candida* in man and animals are attributed to *Candida albicans*. This fungus is found among the normal flora in the mouth, digestive tract, and vagina of perfectly healthy people but under some circumstances and for reasons mostly unknown it may break out and cause severe and even fatal infections, with lesions and eruptions of the skin, nails, mouth, bronchial tubes, and lungs. The source of these outbreaks is difficult to pinpoint because the fungus is so generally present on and within the body of healthy individuals. The presence of special parasitic races is suggested by the fact that some of these *Candida* infections are contagious. Epidemics occur. That predisposition may play a part is suggested by the fact that these infections are said to be more common among workers in certain occupations than in the population in general. Oral infection, known as thrush, is relatively common. Sporadic outbreaks of candidiasis occur in flocks of poultry, the fungus infecting the lining of the crop, stomach, and intestines, sometimes with a high mortality. So far as I am aware, proven parasitic strains of *C. albicans* have not been isolated from plant materials in nature.

Aspergillus Fumigatus and Aspergillosis

The genus *Aspergillus* contains somewhere between 150 and 500 species, depending on whose authority you accept. Several of the species cause infectious diseases, and of these the most important is *Aspergillus fumigatus*. This is a group species containing ten named species, most of them common in decaying vegetation, especially when it is undergoing microbiological heating. Some of these strains of *A. fumigatus* are thermophilic, adapted to growing at a temperature of 50°–55° C (120°–130° F). In one test oat straw, not a particularly nutritious substrate, was inoculated with spores of *A. fumigatus* and was kept well insulated, and in 38 hours the fungus, by its own metabolic heat, raised the temperature of the straw from the original 25° C (77° F) to 55° C (131° F); the maximum rate of temperature increase in any one-hour period was 2.3° C (4.14° F). Alligators use this microbiological heating to hatch their eggs; they pile up moist plant debris over the nest where they have deposited their eggs, the size of the pile evidently being nicely adjusted to provide just the right balance between heat production and heat loss, just the right range of temperature, to ensure the eggs hatching. One wonders how the dimwitted and prehistoric alligator happened onto this clever dodge — the fact that rapidly growing microbes will heat up the materials in which they were growing was first discovered by man in the late 1800s, and there still are a lot of well-educated professional biologists who do not know this, yet the alligators know it and probably have been making use of it since the days of the dinosaurs.

Aspergillus fumigatus sometimes parasitizes animals, especially birds, infecting mainly the lungs and causing heavy mortality — up to 50 percent in young turkeys and up to 90 percent in young chicks. Heavy losses have been reported from this infection in herring gulls, ostriches, and diving ducks in the wild, and in penguins in zoos. The fungus sometimes invades the embryos of eggs in incubators and kills them, and probably does the same in eggs in nests in the wild. It also invades the uterus of pregnant cattle and grows through the placenta into the fetus, which then dies and is aborted. Two British workers estimated that 68 percent of the cases of bovine abortion that they investigated were due to this sort of infection by *A. fumigatus*. It may be equally important in other regions, but unrecognized; to detect it requires someone who not only is aware of the possibility of this fungus being involved, but also has the facilities and know-how to detect the fungus.

That this sometimes parasitic fungus may be with us more commonly than is generally realized is indicated by two papers appearing in a German medical journal, *Deutsche Medizinische Wochenschrift*, in 1974. One (88a) describes a chronic lung infection in, of all people, a surgeon active in a clinic for lung diseases! As determined by exposing culture dishes 20 to 30 cm away from his mouth, then incubating them until colonies of the fungus developed, with every forceful expiration he was expelling more than 100 germinable conidia of *A. fumigatus*. The author of the paper recommends that medical personnel be checked by "cough and speech" plates to avoid their spreading one of the infectious agents against which they are trying to protect their patients. This seems like a reasonable precaution.

The other paper (88b) records severe *A. fumigatus* infection in 15 patients, all of whom died, over a period of four years. In nine of the patients the *A. fumigatus* infection was detected only after death. Most of the 15 patients had some underlying disease that possibly lowered their resistance to infection by the fungus or were receiving medication that might have lowered their resistance. Another possibility, of course, is that the infection by *A. fumigatus* had lowered their resistance to the other infective agents involved in the underlying diseases; they had no evidence on which came first.

Austwick (79) says that *Aspergillus fumigatus* is second only to *Histoplasma capsulatum*, *Coccidioides immitis*, and *Blastomyces dermatitidis* as a cause of systemic disease in man, and there may be many cases where it is present but is not detected.

Blastomyces and Blastomycosis

There are, supposedly, two species of fungus that cause blastomycosis — *Blastomyces dermatitidis*, which causes the North American variety, and *Blastomyces brasiliensis*, which causes the South American. This fungus, or these fungi, occur naturally in some soils, especially soil in animal habitats. It must be generally present in some regions, since it is said that infection in dogs is widespread in Kentucky and Arkansas. Infection is rare in animals other than dogs, only six cases having been reported in cats, one in a horse, and one in a sea lion. Infection in man presumably comes from spores or mycelium in the soil, and any part of the body may be invaded. Infections usually are first detected as skin lesions; the lesion may remain localized or may gradually enlarge. In some cases the fungus may become disseminated throughout the body, resulting in extensive ulceration. Males are infected much more frequently than females — in some studies the ratio has been 15:1. There is no effective treatment.

8

Decay of Wood in Trees and Buildings

In nature, wood, like most other plant and animal materials, is continually being recycled, the wood mainly by fire, wood-inhabiting insects, marine borers and gnawers, and wood-rotting fungi, and of these agents of destruction or recycling the wood-rotting fungi probably are more important than all others combined. The present chapter aims to summarize for you some of the aspects of this decay that is going on about us all the time but of which few of us are aware.

First, something about the chemical and physical makeup of wood. Chemically wood consists primarily of cellulose and lignin, the cellulose accounting for about 55 percent of the total, and the lignin for 25 percent, the remaining 20 percent being composed of hemicelluloses, various kinds of sugars, chemicals of various degrees of complexity that give color and sometimes decay resistance to the heartwood, and around 3 percent minerals. The mineral fraction of 3 percent comes from the soil, and when wood is decayed these minerals are returned to the soil; the other 97 percent is built up from water and carbon dioxide, by means of energy from the sun, and when wood is decayed this portion is broken down into water and carbon dioxide again, with a release of the same amount of energy as was used in its formation.

The cellulose and lignin that make up the bulk of the wood are organized into cells and tissues, with a pattern of arrangement more or less characteristic of each kind of wood; it is the pattern of arrangement of tissues that gives ''grain'' to the wood. The microscopic structure of wood — the kinds of cells of which it is composed, the pits or openings in the walls that permitted protoplasmic connection between adjacent cells when these were alive, the ornamentation within the cells, and the way the cells are arranged into tissues — is even more characteristic of different kinds of wood than is the gross or macroscopic structure. One familiar with them can identify the species of wood by examining microscopically a thin section cut from them. The conifers or cone-bearing trees have a relatively simple structure, the wood consisting of only a few cell types — longitudinally arranged tracheids with connecting bordered pits between them, which constitute about 95 percent of the wood, and small rays that make up 5 percent of the wood. The hardwoods or deciduous or broad-leaved trees have a somewhat more complex structure, but basically the tissues are built up mainly of vessels, fibers, and ray cells.

Cross sections indicate that the wood consists almost entirely of cell walls, the cell cavities being mostly empty. The specific gravity of the wood substance itself, the cell wall material, is approximately the same in all woods, about 1.52, or half again as heavy as water; the density of a given kind of wood depends on the thickness of the cell walls — light woods have thin cell walls and heavy woods have thick cell walls.

How Wood-Rotting Fungi Decay Wood

Wood-rotting fungi digest their food the same way that we digest ours — by means of acids and enzymes — and convert the digested food into some of the same end products, chiefly water, carbon dioxide, and heat. They happen to be able to digest and live a full life on a diet of wood, which of course we cannot do. Different kinds of wood-rotting fungi secrete different groups of enzymes and cause different kinds of decay which can be grouped, according to their gross appearance, into white rots, white pocket rots, brown stringy rots, brown cubical rots, and so on. Some wood-rotting fungi consume nearly 100 percent of the wood they decay and leave only wasted cell fragments behind; others selectively consume certain constituents of the cell walls. Many wood-rotting fungi cause brown cubical rot, so called because in the final stages of decay the wood is brown and, from shrinkage, has become divided into roughly cubical portions, such as occur in mud when it dries in the sun. (See figure 8-1.) The wood decayed by one of these

Figure 8-1. Brown cubical heartrot in a living northern white cedar tree

fungi that cause brown cubical rot may be so thoroughly rotted that it can be reduced to powder merely by rubbing it between the fingers, yet its microscopic structure appears to be unchanged; the cell walls appear to be intact, with no pitting or corrosion or bore holes or other evidence of decay; the cellulose in the cell walls has been consumed in toto, and most of the lignin has been consumed, too, since the weight of the decayed wood is only a small fraction of the weight of the sound wood, but enough lignin remains to give the cell walls the appearance of being sound.

Many thousands of kinds of fungi can grow to some extent on wood, but relatively few — I would judge no more than a couple of thousand species at most — are especially adapted to live on, consume, and decay wood. For these, wood is their preferred diet, and for many of them it is their only diet, and probably has been their only diet for ages.

Most wood-rotting fungi are in a special taxonomic group — the *Polyporales* — which includes the pore fungi and some of their relatives. Fruit bodies of *P. rheades* are shown in figure 8-2. There are wood-rotting fungi in other taxonomic groups also, notably among the gill fungi or *Agaricales*, but by far the greater amount of decay of wood, both in trees and in wood in use, is caused by members of the *Polyporales*, and by far the greater number of *Polyporales* live on, consume, and decay wood.

Some of these wood-rotting fungi are generalists, to the extent that they can grow on and decay the wood of many different species of trees, and others are specialists, growing only in the wood of one or a few closely related species of trees. A few will be described.

Fomes pinicola decays the heartwood of living and of recently dead trees of more than 100 different species. (*Fomes* comes from the same root as *foment*, to heat up or burn; in the days before matches, fire often was held over for hours or days in a dried fruit body of any one of several species of *Fomes*; these fruit bodies when ignited do not burn with a flame, but glow, and when such a glowing fruit body is covered with ashes it will maintain the fire for a long time. The specific name *pinicola* means "inhabiting pine.") *F. pinicola* occurs much more commonly in coniferous trees than in hardwoods, but is found at least occasionally in hardwood trees, stumps, and logs. It has a very wide

Figure 8-2. Fruit bodies of *Polyporus rheades* on a living oak tree, the heartwood of which is being decayed by this fungus

geographic range, and the fruit bodies found on different species of conifers around the world in the northern hemisphere, and from Alaska to the mountains of central and southern Mexico, look surprisingly alike. It does not decay wood in use.

Polyporus sulphureus grows in and decays the wood of hardwood trees primarily, but it is found on conifers occasionally. Like *Fomes pinicola* it has been recorded in or on more than a hundred species of trees and, also like *F. pinicola*, it has an exceedingly wide geographic range. It also does not decay wood in use. There are other wide-ranging wood-rotting fungi, both in the kinds of trees they inhabit and decay, and geographically, but these examples will serve to illustrate the fact that while these fungi are specialized to the extent that they live only in wood, within that environment they are exceedingly unspecialized.

Polyporus robiniophilus, on the other hand, causes a decay of the heartwood of locust trees only, and it is relatively rare even there. *Polyporus amarus* inhabits the heartwood of incense cedars only, and *Polyporus basilaris* causes a rot of the basal portion of the trunk of Monterey cypress only and never has been found outside of California, as is true of its host also. The host trees of these three species have wood that is highly resistant to decay, by virtue of compounds deposited in the heartwood as this forms. As a general rule, trees with heartwood highly resistant to decay are subject to attack by very few kinds of wood-rotting fungi, at least so long as the trees are alive. Conversely, trees with heartwood highly susceptible to decay are likely to be attacked by many species of heart-rotting fungi. This does not always hold, however. Our common aspen *Populus tremuloides* has no heartwood distinguishable as such; the wood of the inner portion of the trunk is of the same color as that of the outer portion, the sapwood. This "heartwood," if it is heartwood, is susceptible to decay by a great variety of fungi, yet in living aspen trees one seldom finds any fungus but *Fomes igniarius* causing heartrot. (See figure 8-3.) Very probably *F. igniarius* does so well in aspen trees that it is able to overcome its competitors. Wind-thrown aspen trees or logs or bolts may be attacked by a multitude of wood-rotting fungi to the almost total exclusion of *F. igniarius*.

All the fungi discussed above cause decay of the heartwood of living trees or of recently dead trees, either standing or fallen. Other wood-rotting fungi grow only in the recently dead branches of a given size of a given kind of tree and at a given time interval after the branch has died. Most of the fungi that cause decay in living trees do not continue to grow much after the tree has fallen, and almost none of them cause decay of wood once it has been sawn or otherwise processed into timbers or lumber. Once a tree, of whatever species, has fallen, it is not likely to be decayed into dust by one fungus alone, but rather it is attacked by a succession of different fungi, each following its allotted predecessor in this ecological parade, and in turn being succeeded by its allotted successor, in a fairly well-established and relatively unalterable order. Precedence and protocol prevail even in death and dissolution.

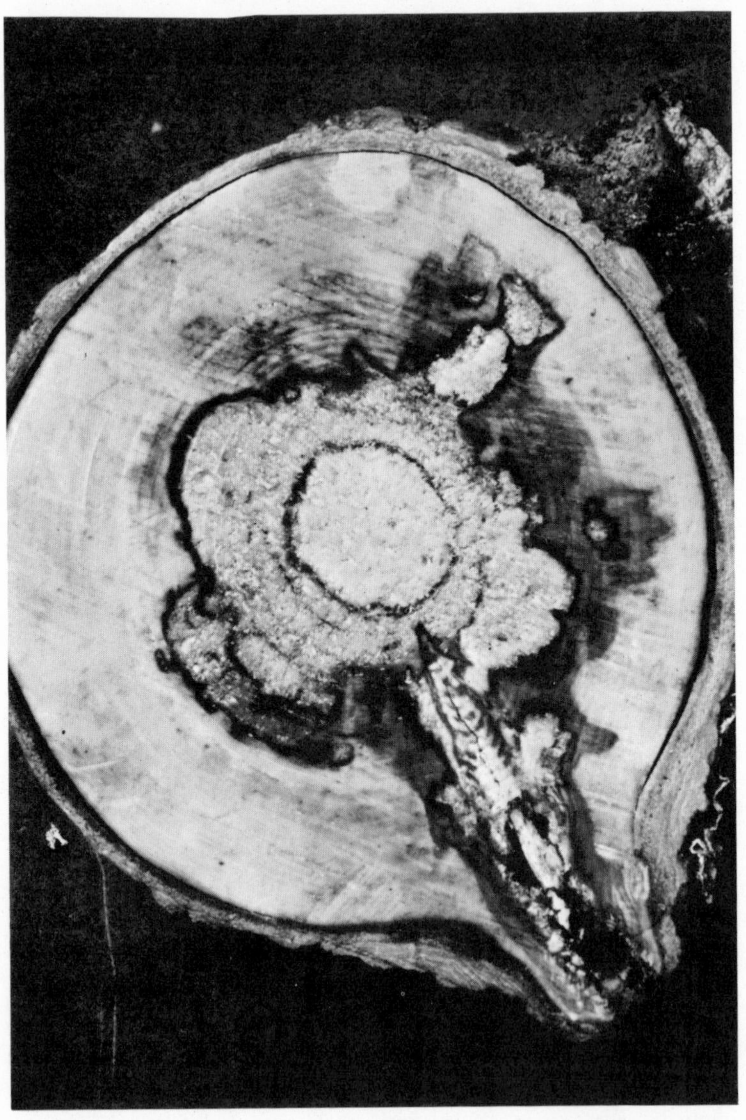

Figure 8-3. Cross section of an aspen tree decayed by *Fomes igniarius*

The first sound work on decay of wood was published in 1878, by Robert Hartig, a forester and professor of botany at the Prussian Royal Forest Academy, Eberswalde, Germany. His book, *Zersetsungserscheinungen der Nadelholzbäume und der Eiche*, is still respected (but, like many classics, probably is not read very much — it isn't exactly easy reading, by any means). It included beautiful drawings of the microscopic features of the different sorts of decay that he described. The state of investigation of decay of wood up to that time Hartig summarizes in a single sentence: "To previous works of other investigators I fortunately needed to pay little heed, since such were almost completely lacking." This in spite of the fact that since the 1830s, at least, some wood in use, such as railroad ties and bridge timbers, had been protected from decay by impregnating it with creosote; that is, the disease was cured before its cause was known, as happens occasionally even today.

Since Hartig's time (he died in 1901) a good deal of study has been devoted to various aspects of wood decay, and a good many books have been written on the subject. Nevertheless, the general public know very little about it, and continue to be taken in by false doctrine; and architects continue to design and builders continue to build houses that are subject to damage and destruction by wood-rotting fungi, sometimes within a few years after they have been constructed. Plant pathologists frequently are asked, for example, whether removal of a fruit body of a wood-rotting fungus from a tree, or a fence, or a windowframe of a house will stop the decay in the tree or fence or windowframe (it won't, of course), a question that betrays a total ignorance of the decay process. Aside from those few exposed to some forest pathology or plant pathology, people in general, even those who are otherwise well educated and some of whom almost daily solve the world's problems, have no more understanding of the nature and cause of wood decay than did the average unlettered peasant in medieval times, and even some plant pathologists with a Ph.D. degree have some rather curious and foggy notions about it.

There are, in general, two problems of decay in wood: (1) that in living trees; (2) that in wood in use. Each will be discussed separately.

Decay in Trees

All trees are subject to decay, both before and after they die. In living trees this decay, at first limited to the heartwood but later often progressing out into and through the sapwood, may limit the length of life of individual trees and may be the determining factor in the longevity of certain tree species. Aspen and balsam fir, for example, throughout most of their extensive geographic range, are short-lived because they are at an early age invaded by wood-rotting fungi — the aspen primarily by *Fomes igniarius*, which enters through branch stubs, grows up and down the interior of the trunk, and decays it, and the balsam fir by fungi that enter the roots, decay their interior, and progress upward into the basal portion of the trunk, weakening the tree so that it eventually falls over. Apple trees in the midwestern United States seldom reach an age of more than 35 years because of wood-rotting fungi that decay the trunk and branches, and many of them are extensively decayed before they are 25 years old. In New York State apple trees may live and bear fruit for 50–75 years or more, and in Europe some of them live even longer. Within a given species of tree the damage from heartrot may vary greatly within a relatively small area: In some locations in northern Minnesota the common aspen tree struggles into a crooked and crummy-looking travesty of a forest tree, riddled with decay by the time it is 50 years old, whereas a hundred miles away there may be a stand where all the trees grow tall and strong, with no interior decay or outward blemish, and form what, for aspen, are proud forest giants.

The long-lived species of trees, such as bristlecone pine, redwoods, sequoia, western red cedar, incense cedar, bald cypress, white oaks, black locust, to name some representative species, are highly resistant to heartrot, by virtue of materials that are accumulated and deposited in their heartwood as it is formed and that are toxic to fungi. It is, indeed, chiefly these deposits that give the heartwood its characteristic color. The heartwood of decay-resistant trees is, by nature, impregnated with wood preservatives. They are resistant to decay, but not immune from it; given sufficient time and conditions favorable for their growth, wood-rotting fungi will consume even these decay-resistant woods. A cellar door, for example, constantly exposed to the weather, will last a good deal longer if it is made of redwood than if it is made of decay-susceptible wood that has not been treated with a wood preservative. If the cellar door is made of redwood and has moreover been given several liberal brush treatments with pentachlorophenol, it may not be permanent, but neither is it going to fail very soon from decay.

Not all dark-colored woods are highly decay resistant. Philippine mahogany has at least a fairly dark color, but it is moderately susceptible to decay, as some people found out, to their cost, who assumed that because of its name (it is not mahogany at all, nor is it related to the true mahogany) and its color, it would be suitable for use in high-decay-hazard places. They imported it for use as mine timbers, which failed quickly from decay.

Very heavy and dense woods are, by virtue of their density, more resistant to decay than are lighter woods. In the first place, there is more wood material there, and loss of some of it is not so serious as it is with lighter woods; second, in the denser woods there is a very narrow margin between not enough water for decay to occur and too much water, about which more will be said in the section on decay of wood in use.

Heartrot Hazard in Trees Planted Out of Their Natural Range

Many kinds of trees, like many other kinds of plants and like many kinds of animals, are closely adapted to the region where they occur naturally. In any community the different species exist in a fluctuating and often precarious balance, dependent upon one another in many subtle ways that we still know nothing about, and they have intimate ties with the living and nonliving things in the soil, and with all the many and complex things that go to make up climate; this is not at all strange, since the members of these communities and their precursors have been evolving together for hundreds of millions of years. Because man can live almost anywhere, he is inclined to endow other living things, including trees, with this same adaptability. Actually, many varieties of man have become adapted to this or that relatively restricted environment on earth; this was even more true a few hundred years ago, before the days of great explorations, than it is now, because in the meantime many of the special varieties of man that had developed here and there have been largely eliminated by diseases and other accompaniments of civilization introduced to them by travelers. Man has always been a wanderer to some extent; otherwise he could not have covered the globe as he has, with a single species. Trees could not wander, except in the sense of their seeds being carried by wind or animals or other means. Within any one species of tree that is widely distributed geographically there are different varieties that have developed by infinitely slow adaptation to the soil, climate, and company of the locality where they occur. Some of these varieties are so delicately adapted to the environment in which they occur that if they are moved — that is, if their seeds are planted — more than 50–100 miles away from where those particular individuals were growing, the resulting trees may be subject to damage by many things, including heartrot, that never bothered them before. A few examples will make this clear.

In the 1600s European foresters began to establish plantations of forest trees. There are only three species of native timber trees in Europe — Scotch pine, Norway spruce, and European larch — and none of them are exactly superior timber trees, at least compared with some of the magnificent timber trees of western North America. Once these superb forest trees of western North America became known to the European foresters, they envisioned plantations in Europe that would rival the forests of the West. After all, why not? So seeds of many of these western conifers were taken to Europe, seedlings were grown, and plantations were established. None of these plantations were really successful, and many of them were dismal failures. In many of them, the young trees, barely out of their teens, were plagued by a host of minor and major pests, among them heartrots. Douglas fir, for example, which in western North America still is a relatively youthful and fast-growing tree when it is 300 years old, and which may live to be 600 years old before it is damaged much by heartrot, in the European plantations was attacked, crippled, and often destroyed by heart-rotting fungi before it was 50 years old. The fate of most of the other introduced trees was largely similar.

It was not so much a matter of the trees being exposed to a new and more deadly array of parasites and pathogens as it was a matter of the trees simply not being suited to the new environment, or the environment not being suited to the tree, whichever you like. This turned out to be not an isolated case, but the usual thing, the rule. Jack pine (*Pinus banksiana*) has a very large range, geographically, occurring from central Minnesota north and west to Alaska, and eastward to New England and through the southern half of Canada. Throughout this relatively vast range the trees look very much alike; they suffer from the usual array of insect pests and fungus diseases but, by and large, not much more so in one place than another. Jack pine once was considered to be only a weed tree, of little value, but this has changed, and it is being widely planted, by private pulp-and-paper companies among others, as a valuable source of wood products. Some of these plantations have been established without regard to source of seed and some of these are turning out to be dramatic failures. "Source of seed" plots have been established at the Forest Experiment Station of the University of Minnesota at Cloquet. Seeds were gathered from different places throughout the range of jack pine and planted in separate blocks. Even while the trees are still young very apparent differences have developed among those that come from different sources, especially in rate of growth, form of growth, and resistance to damage by insects and disease. The trees still are too young to have been much injured by heartrot, but as they grow older it is certain that those from some sources will be much more injured by heart-rotting fungi than those from other sources. Nor are the best ones necessarily those that came from closest to where the plots were established. The moral of this is that, if you want to establish productive plantations of jack pine in different regions, you had better accumulate some information by means of "source of seed" plantings on what geographic varieties of jack pine are likely to do best where. The point is that all trees are subject to heartrot, or decay of the heartwood, and those grown far from their natural haunts are likely to be especially subject to damage by heartrot, and at an early age. They are subject to damage by many other pests also.

As always, there are exceptions to the general rule that trees taken far from their native haunts are doomed to a short and miserable existence. Three notable exceptions come to mind.

Monterey pine occurs naturally in a restricted area on the coast and nearby islands of central California and northern Baja California. In its native range it is just another somewhat scrubby pine, of little or no commercial value, although it has been grown as an ornamental or decorative tree up and down the Pacific coast. In spite of its unprepossessing appearance in its home range, it was carried to other countries to see how it might do, and in some of these countries it did wonderfully well. In such widely separated places as New Zealand, Australia, and South Africa it grows rapidly and produces saw timber in a relatively short time — in New South Wales trees at 35 years of age averaged 50 feet in height and 18 inches in diameter, very rapid growth indeed for any kind of pine anywhere. Why? If it was so good, how come it was so limited in its natural range and relatively unsuccessful even there — it supposedly had the same chance to develop that its many competitors had, from way back when conifers started.

The case is somewhat different for Eucalyptus, of which there are many species native to Australia and to neighboring islands. Some of these are excellent timber trees, and a few species have been carried to South America, to Mexico, and to California, and in some areas of those regions some species of Eucalyptus do very well, as either ornamental or timber trees, or both. It even thrives as a street tree in Mexico City, at an altitude of 7300 feet and with the air so polluted it chokes you, and near Bogotá at an altitude over 9000 feet and in a climate quite different from that of Australia. However, to illustrate the uncertainties of naturalizing even so widely adapted a tree: One entrepreneur in California established extensive plantations of Eucalyptus, from which, when they matured as timber trees, he expected to clear millions of dollars. The trees grew bravely for some time, and reached a height of 40 to 50 feet and a diameter of 8 to 10 inches. Then, one winter, unusually cold weather descended on that part of California for a time and, as a result, all of his Eucalyptus trees perished.

The Ginkgo tree (*Ginkgo biloba*) is another odd one. There is only one species in the genus and it is, in fact, the sole surviving member of its particular family and order. It is related to the Cycads, the pre-coniferous trees that predominated in the Carboniferous Age and of which a few members still survive in tropical regions. At the time dinosaurs roamed the earth, the Ginkgo tree was worldwide in distribution. The Ginkgo tree no longer survives in the wild, nor has it for thousands of years, but it has been preserved in plantings around monasteries in China and Japan, where it originally grew. It was carried to the Western Hemisphere first as a botanical curiosity, but later was found to be so well adapted there that it became at least fairly widely planted as a shade and ornamental tree. It survives even as a street tree in New York City in spite of the poisonous air and thrives in the far from balmy climate of Minnesota. Moreover it has almost no really seriously damaging insect pests or fungus diseases, and it certainly is not subject to early death from heartrot.

How Heart-Rotting Fungi Enter Trees

Most fungi that decay the heartwood of living trees can enter the tree only through exposed heartwood. As trees grow, the lower branches may be shaded out, and die. If these branches are large enough to have heartwood, this will be connected with the heartwood in the trunk of the tree. Such branches will be decayed by various fungi, and eventually they fall off and expose the heartwood, through which the heart-rotting fungi will enter. Often when branches are trimmed from trees, heartwood is exposed. The wounds so made can be covered with various kinds of wound paints or wound dressings, which improves their appearance but does not in any way prevent the entrance of decay fungi. Some "tree care" people will dispute this, but without any basis other than that since they are in the tree-care business, anything they do must have some beneficial effect on the health of the tree.

Foresters very frequently bore into trees in the woods with a tool called an increment borer. With this they extract a core about half the diameter of a pencil and of a length equal to the distance that the borer penetrated into the tree — say eight or ten inches. From subsequent study of these cores such things as age and rate of growth of the trees are determined. Decay fungi enter through the holes made by these increment borers, of course. In the 1930s a forest pathologist interested in determining whether such decay could be prevented bored into a variety of hardwood trees with a standard increment borer and, as soon as the borer was withdrawn from the tree, swabbed the hole with a disinfectant, then applied various wound dressings — some of them containing really potent and proven fungicides — to the interior wall of the hole. After some of the holes were swabbed with the disinfectant, they were plugged by pounding into them tightly fitting solid cylinders of wood that also were impregnated with a potent fungicide. The controls were left untreated.

About ten years later these trees were felled and split to expose the wood traversed by the original increment borer. Heartrot, originating at the increment borer hole and extending as much as a foot or two up and down from it, was present in every tree. There was no difference between the untreated controls and those treated with any of the fungicides. The fungicidal treatment did not prevent decay fungi from entering the tree nor did filling the hole with fungicide-treated wood slow down the progress of decay. In these tests, the fungicidal treatment was applied within seconds after the increment borer was withdrawn from the tree — but that evidently was long enough to permit the fungi to enter. Enter by what means? We don't know. But we do know that the treatments used were applied faster and the compounds used were better than those applied after branches are removed from shade trees.

Frostcracks, firewounds, blazes, or any other cuts, hacks, or borings that expose the heartwood permit heart-rotting fungi to enter a tree trunk. Some trees, in forests where logging is carried on, reproduce mainly by sprouts that grow from a recently cut stump or from the roots of the original tree. The heartwood of the stump is quickly invaded and decayed by heart-rotting fungi, and, if heartwood forms in the sprout and joins that of the stump before the fungus dies, it will be invaded from that source. In southern pine forests heartrot caused by *Fomes annosus* (the fungus happens to be a real killer, as some heart-rotting fungi are, especially in plantations, as distinguished from the primeval forest) spreads through root grafts from the stump of a cut tree to adjacent trees. Some heart-rotting fungi live in the soil, evidently as components of the normal soil flora, and they may enter the young and presumably healthy roots some distance out from the trunk of the tree, gradually decay these roots, and progress toward the trunk. By the time the decay has gone more than a short distance up the interior of the trunk the tree is likely to fall over.

Progress of Heartrot

Once heart-rotting fungi have invaded and become established in a tree, they gradually progress through the tree longitudinally. The rate at which this decay advances depends on the tree, the fungus, and perhaps the location. *Polyporus amarus*, one of the few heart-rotting fungi that cause decaying in living incense cedar, a highly decay-resistant species, may advance at the rate of only 0.01 to 0.35 foot per year. One one-hundredth of a foot is about an eighth of an inch per year, or expressed another way, a foot in 100 years. *Fomes igniarius*, a more rapidly growing fungus, growing in a more decay-susceptible species of tree, advances at a maximum rate of 1.23 feet per year. The decay progresses outward more slowly, but eventually most or all of the heartwood may be consumed, leaving only a thin layer of sapwood. This does not necessarily doom the tree to an early death by windthrow. Such hollow trees are more subject to be blown down by a hurricane than are perfectly sound ones, but a hollow cylinder is almost as strong as a solid one — the hollow stems of grasses such as wheat and rye, or bamboo, can support a relatively tremendous load — and so hollow trees may continue to live for decades or even centuries. Some of them harbor such delightful fellow creatures as owls, hawks, and flying squirrels and so add variety to the landscape, and for that reason alone should be cherished.

After the fungus that is causing the heartrot in a given tree has been growing in that tree for years, decades, or centuries, it may form a fruit body or sporophore (or several or many of them) on the outside of the tree — usually at a branch stub or wound that exposes the heartwood. The function of these fruit bodies is to produce spores, which serve to disseminate the fungus, as seeds serve to propagate higher plants. Depending on the species of fungus, these fruit bodies may be annual, new ones being formed each year, or they may be perennial, a new spore-bearing layer being added each year for as long as the fruit body remains alive, which ordinarily is as long as heartwood remains available for it to decay, and may be as long as 50–75 years. You should be able to see by now why removal of such a fruit body from the outside of the tree will not affect the progress of decay within the tree.

Some fungi that cause heartrot in living trees seldom form fruit bodies. This is especially true of those that decay the more decay-resistant trees, the redwoods, cedars, and cypresses. Heartrot caused by *Polyporus amarus* is common in incense cedar, but few fruit bodies of this fungus have ever been collected. A brown cubical heartrot is common in northern white cedar in Minnesota, but although several of us have for years been on the lookout for fruit bodies of the fungus that causes this rot we have yet to see a single one; perhaps it fruits on other kinds of trees, but if it does we have yet to discover this.

Control of Heartrot

There isn't really much that can be done to control heartrot in trees in the forest, and even less that can be done to control it in trees in the backyard or on the estate. In forest trees the only control measures are indirect — ecological: (1) Plant the trees, or encourage natural reproduction, on sites favorable to the species, where experience has shown that the trees do well and live a long and healthy life. Experience of this sort is accumulated slowly. (2) Select varieties or biotypes or variants that do best in a given region. This also is a longtime business, involving the collection of seeds from many different areas over the range of the species in question, and planting those from each area in plantations in different test sites. Of course replicate tests, in time and space, are needed to really be sure, which means that a few hundred years may go by before the results are in. However, the few hundred years go by in any case, and, if no such tests have been begun, after the lapse of a couple of centuries we still would not have the information we need. (3) Harvest the trees before losses from heartrot become serious. This is the basic, direct approach that has become widely adopted, especially since some of the modern processing techniques permit use of practically the entire tree — bark, branches, twigs, and even partially decayed wood.

Theoretically some of the methods cited above could be applied to ornamental trees. It is possible to plant long-lived, decay-resistant species such as, say, white oaks, but it might require a century or more for them to grow up to a respectable size, and relatively few people are willing to plant for that far in the future. Native, adapted trees could be selected in preference to exotic, very short-lived, nonadapted trees — if you can get them; but most people are not likely to give this much consideration: somehow they seem to believe that all trees will live forever, and that, if the trees on their property languish and die, it is due to some specific insect or disease that can be cured by application of a suitable insecticide or fungicide.

Heartrot often is found in old, established trees that are prized for their ornamental or their sentimental value. In the past it was a common practice for well-to-do owners of such trees to have a tree surgeon remove the decayed wood and fill the resulting cavity. Orginally concrete was used for this purpose and from time to time when the tree eventually died and fell over or was removed, this pillar of concrete remained as a mute memorial to the tree surgeon. Later various other materials, usually combinations of asphalt with sawdust or some other diluent, came to be used. Without going into all the details about this, it may be said that it is virtually impossible to eliminate all the decay-causing fungus or fungi from a given decayed tree, and it also is impossible to prevent wood-rotting fungi from establishing themselves in the wood exposed when the cavity was made — recall the account, above, in which wood-rotting fungi became established in the wood next to increment borer holes in spite of all the precautions taken to keep them out. There has never been any evidence that the progress of decay in trees was arrested, or that the length of life of the trees was in any way prolonged, by this sort of cavity work; it has been largely discontinued by reputable tree-care men, which is just as well.

Decay of Wood in Use

Wood protected from fire, fungi, and attack by insects is exceedingly durable; time alone causes no appreciable deterioration in wood, at least within a span of thousands of years. Under conditions favorable to the development of decay-causing fungi, however, the effective life of wood may be no more than a few months. How long wooden houses or wooden boats or wooden posts and poles will last depends upon how well they are protected from decay (in many areas they must also be protected against damage by termites). For good protection against the ravages of decay, it helps to know something of the conditions that permit wood-rotting fungi to grow, the conditions necessary for decay to occur.

Wood-decay fungi need the same things that other fungi need and that most other forms of life also need, namely food, water, oxygen, and a favorable temperature. Their food is furnished by the wood that they decay. As indicated above, the wood of different kinds of trees differs greatly in resistance to decay, the darker colored and heavier woods being, in general, more resistant to decay than the lighter colored, light-weight woods. This difference in decay resistance is a relative thing, a matter of degree; no wood is immune from decay.

All wood-rotting fungi require, for their growth, some free water; none of them can make do with a moisture content in equilibrium with a relative humidity of 70–80 percent, as can some of the fungi that invade stored products. When there is just enough water in wood to saturate the cell walls, but no free, liquid water is present in the cell cavities, the water content is said to be at the "fiber saturation point." This is a moisture content of 25–28 percent of the oven dry weight of the wood. This is the lowest moisture content at which decay can occur, and relatively few wood-rotting fungi can cause much or rapid decay at this moisture content. To grow well, they need a supply of free water, a moisture content somewhat above that of the fiber saturation point. At least one wood-rotting fungus, *Merulius lachrymans*, to be described in some detail shortly, once it is established in wood in a fairly humid environment, apparently can supply part of its moisture requirements from its own metabolic water, but most wood-rotting fungi need moisture from another source. One way to protect wood from decay, then, is to keep it dry, and that is the principal decay-preventive method used in houses and other wooden buildings. When for one reason or another this protection fails and the wood becomes wet, decay almost inevitably follows.

An excess of water will also prevent decay in wood. Wood pilings driven down in the mud of river bottoms or in shallow lakes have been recovered more than a century later, not only sound but also with a more attractive grain, when sawed into lumber, than freshly cut timber. Pine logs that sank in midwestern rivers during the log drives of the 1890s were recovered 40 to 50 years later in sound condition and made excellent lumber. Wood submerged in sea water will not decay, but if not impregnated with a preservative, and sometimes even when impregnated with a preservative, it will be consumed by marine borers and gnawers. It is assumed that the absence of decay in wood submerged in water is due to lack of oxygen, a reasonable assumption, to be sure, but an assumption still. It could be due simply to an excess of water diluting the enzymes and acids of the wood-rotting fungi to such an extent that they no longer can digest the wood.

Another moisture relationship deserves to be mentioned briefly here. Although no wood-rotting fungi can grow in and decay dry wood, some of them, once they have become established in the wood, can endure drying for years without losing their vitality, and if the wood again becomes moist they can revive and begin growing within hours. Intermittent wetting of wooden members of a house or other structure, especially if followed by moderately slow drying, will result in decay unless the wood is otherwise protected, as by having been impregnated with a preservative.

In their growth at different temperatures the wood-rotting fungi are pretty much standard middle-of-the-road conservatives. They grow best at a temperature around 30° C (86° F), cease growing or grow very slowly at a temperature of about 5° C (41° F), and are rendered inactive or killed at any temperature much above 40° C (104° F). None of them grow at temperatures below freezing as do the snowmolds and as do some of the fungi that occur on refrigerated food products, and none of them are adapted to a temperature of 53°–55° C (125°–130° F) as are some of the thermophilic fungi that are responsible for the microbiological heating in hay, stored grains, and piles of wood chips.

In practice, the important limiting factor for decay of wood in use is moisture. If the wood is moist enough long enough it will decay, and sometimes this is not very long. Under conditions exceptionally favorable for them, wood-rotting fungi have been known to cause serious damage in wooden buildings, including houses, while these were still in the process of construction. Wooden fence posts five inches in diameter and of decay-susceptible wood not treated with a preservative may decay to the point of failure in three short summers in northern Minnesota, where the growing season for wood-rotting fungi is only three months long. (See figure 8-4.)

Given the conditions that permit or promote decay, for decay to occur there must be present some inoculum of wood-rotting fungi, in the form of spores or fragments of mycelium. Judging from various sorts of evidence, inoculum of wood-rotting fungi is just about as universally present as anything can be. There is no record, anywhere, of wood having been exposed to conditions that promote decay without decay having developed. As one example, during World War II, green lumber cut from trees on the West Coast of the United States was sometimes made into large bundles and these were surrounded and tightly bound by steel bands, loaded into ships, and carried via the Panama Canal to the East Coast. The wood throughout some of these bundles was fairly well decayed by the time the bundles arrived at their destination. All these boards, in other words, between the time they were cut from the logs and the time they were bound with steel straps into tight bundles, a matter of hours at most, must have been thoroughly inoculated with spores or mycelium of wood-rotting fungi.

Another example: Once we put small, moist blocks of aspen wood, which is susceptible to decay by many kinds of fungi, into screw-capped jars, autoclaved them to kill any fungi that might be present, then, at various times, exposed several of the blocks to outside air for an hour, put them back in the jars, sealed these, and kept them in the laboratory to see what would happen. Decay developed in all blocks so exposed, even those exposed in midwinter in St. Paul — decay that could have come only from inoculum of wood-rotting fungi in the air at the time. It is questionable whether lack of inoculum ever is a limiting factor of decay in wood anywhere, either in trees or in wood in use. Given a moisture content and temperature favorable for decay, decay will develop — the inoculum will be there. This may seem obvious, but it evidently is not; as we shall see shortly, when control of house rot is discussed, some have considered, and some still consider, sanitation — avoidance of inoculum — to be very important in avoiding rot. It isn't; there is just too much inoculum of wood-rotting fungi present everywhere at all times. It is highly probable, in fact, that just about all wood

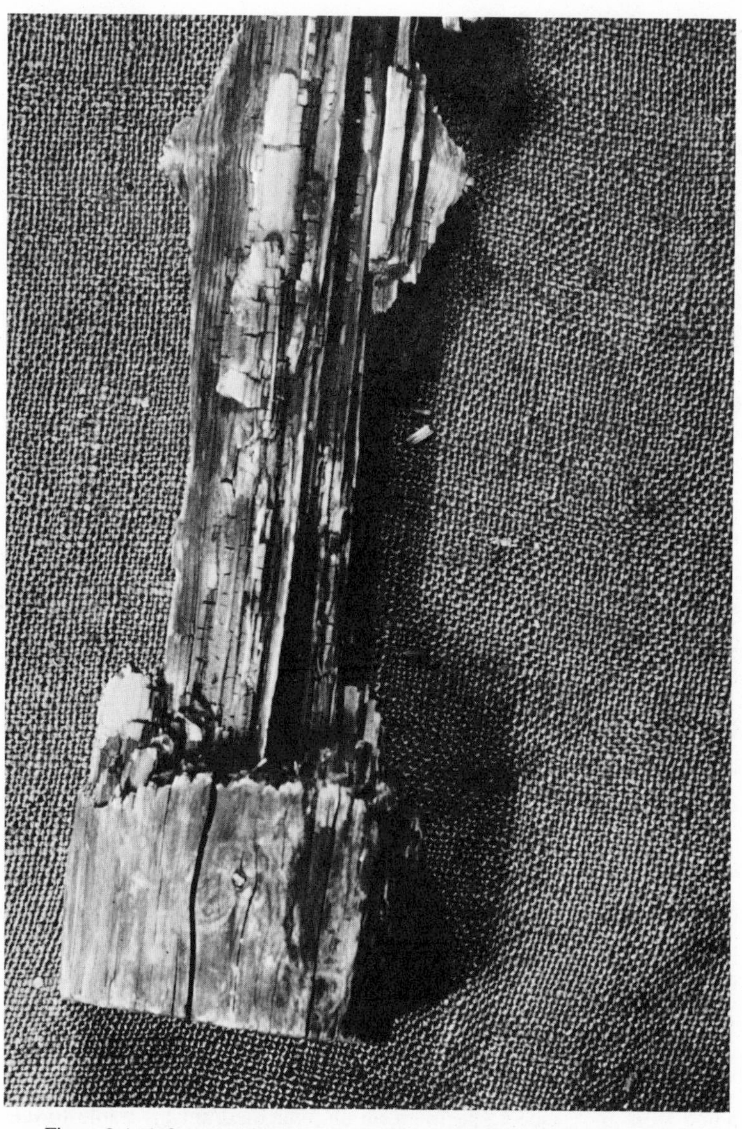

Figure 8-4. A fence post badly decayed after having been in service for three years at Cloquet, Minnesota, where the growing season is only about four months long

is moderately to heavily inoculated with wood-rotting fungi long before it ever gets to the construction site. Even lumberyards have decay problems in their stocks of wood, sometimes rather serious ones.

Effects of Decay

The major effect of decay is loss of strength of the wood. Various chemical and physical changes accompany this — the wood becomes more permeable to liquids, including water; it absorbs water vapor faster than sound wood and loses it more slowly and, at a given relative humidity, retains a higher moisture content than sound wood, so that once decay is under way it tends to be self-perpetuating; also as decay progresses the wood changes in color and decreases in density, since the wood substance itself is being consumed. However, in most wood in use loss of strength is the most important result of decay. Those who deal with the engineering aspects or mechanical properties of wood recognize several different kinds of strength in wood, and have standard machines and procedures for testing them. Some kinds of decay will reduce one strength factor much more than another, and even before decay has become visible to the eye by any alteration in color or texture of the wood, toughness or impact strength may be reduced by 75 percent.

Detection of Decay

In practice, decay in wood seldom is detected until it has progressed to a fairly advanced stage and the wood is discolored or has collapsed, or swatches of mycelium grow over the walls or ceiling or fruit bodies sprout out of the woodwork — a fearsome sight. In regions of high decay hazard in houses and other buildings, professional inspectors or decay surveyors can be hired to examine the structures; with experience they have learned where decay is most likely to occur and they have developed some expertise in detecting decay by prying and probing and boring.

Theoretically decay in wood should be detectable by plating bits of wood from the suspect specimen on an agar medium favorable to the growth of wood-rotting fungi. Most wood-rotting fungi grow readily on common agar media, and many of them can be identified according to genus and species by their so-called cultural characteristics on agar media. In practice this has its limitations. Even if wood is obviously decayed, not all pieces of it put on agar will yield the decay fungus, and sometimes very few pieces will. So if we do not recover the fungus from bits of the suspect wood put on agar, we don't know whether this was because the fungus was not present or because it did not grow out. And if the fungus does grow out on the agar, all this tells us is that the fungus is present in the wood. It is good circumstantial evidence that decay is present, but it does not tell us if enough decay is present to affect the strength of the wood.

Similiarly, microscopic examination of stained sections of wood sometimes is not too helpful in judging whether decay is present. If mycelium is found in such examination, and is associated with ''bore holes'' or other obvious destruction of the cell walls, it is evidence that decay is present. It does not tell whether this is of recent origin or whether it is old — the decay may have been present in the tree from which the lumber was cut. Also since decay can occur without being accompanied by any microscopic changes, if no changes in cell structure of the wood are detected this does not prove the absence of decay. So very often even after a thorough examination we cannot testify that decay is or is not present in a given piece of wood, but only that such and such tests were made and such and such results obtained.

Decay in Houses and Other Buildings

In climates where the air is consistently humid throughout a good portion of the year, decay in houses and in other buildings can become a very serious problem. In the United States this sort of decay is prevalent in the Gulf States, in states bordering on the Atlantic Ocean south of New York, and, to a lesser extent, in the coastal areas of Oregon and Washington. It is not limited to those regions by any means, but is especially prevalent there. House rot is common in Great Britain and all across northern Europe, where the relative humidity is so high so much of the year that once the wood becomes wet it takes a long time to dry out, especially where central heating is lacking or is not used much.

In Great Britain, ''dry rot'' legally means rot caused by a specific fungus, *Merulius lachrymans*, and if a house or other building is sold with the stipulation that it is free from dry rot this means that it is free from rot caused by this fungus. Theoretically, at least, the house could be on the verge of crumbling into the moat from decay by other fungi, and the buyer would have no recourse. The type of decay caused by *M. lachrymans* is not especially characteristic — just a brown cubical rot such as is caused by a dozen other fungi that occasionally decay wood in use. If fruit bodies or masses of mycelium are present on the decayed wood, as they sometimes are in enclosed places, identification is certain; if they are not, then presumably the identity of the fungus causing the decay could be determined by isolating cultures of it on an agar medium.

Merulius lachrymans is said to be rare in the United States, or not to occur at all, presumably because the conditions it needs — a combination of low temperature and high humidity — do not prevail over large areas here. In Great Britain it is said never to have been found in nature, and with so many people there going on fungus forays in the fields and woods, if the fungus were present in nature it should have been encountered at least once in a while. Perhaps it is a species that has developed along with the development of wooden housing, but that seems a bit farfetched.

In any case, the rot it causes can be a fearsome thing, as well as insidious. A house is constructed, and is lived in for several decades with nothing out of the ordinary occurring. One dark and gloomy winter the family take off for a vacation in the sun and close up the house for a month. They return to find thick white fans of fluffy mycelium spreading over the inner wall of the room next to the conservatory, and fruit bodies of *Merulius lachrymans* sprouting out of the wainscoating. There is a smell of decay and dissolution in the air. The dry-rot surveyor comes, opens up the wainscoating, and finds the spaces between the studdings in the interior of the wall filled with gobs of mycelium of *M. lachrymans*, and all the studdings crumbling and collapsing from decay. The decay has advanced into the wood beneath the floor of the room also, and when floorboards are taken up the subflooring and the joists on which the floor rests are found to have been decayed to just about the point of failure, and festoons of mycelium of *M. lachrymans* wave silently in the quiet air beneath. To determine the extent of the decay and to remove the decayed wood and replace it with sound construction is, of course, extremely costly. Sometimes a house in that condition is simply abandoned.

In this case the decay must have been present and progressing for at least some years, and probably for a decade or more; it became suddenly apparent because the house was closed up and unventilated, so that the fungus could grow out on the side of the wall facing the room; ventilation, in this case, would not have prevented the decay, it would only have delayed for a short time its detection. Some wood-rotting fungi decay the wood without producing any mycelium on the surface of the wood, and, indeed, without producing much mycelium, at any one time, within the wood — what mycelium they do form within the wood undergoes autodigestion as the fungus progresses from wood with which it is finished to new wood. They also form few or no fruit bodies to reveal their presence. Not so *Merulius lachrymans*. Given an enclosed space where the humidity remains high, in part from metabolic water produced by the fungus, it can form great masses of mycelium and relatively huge fruit bodies — one has been documented that was five feet long, as big as a small rug, and formed, moreover, on the brick floor of a cellar (91). The other fungi that decay wood in buildings in Great Britain are grouped together as "wet rots"; all of these require water that comes from some source other than their own metabolism, whereas *M. lachrymans*, as noted above, once it is established, evidently can furnish part of its own; this is the only way it could take over entire buildings in the way it sometimes does.

In a survey of 9810 buildings in 11 districts of the United Kingdom, dry rot was found in an average of 13 percent and wet rot in 26 percent of them (91). In some districts more than 40 percent of the buildings examined had some decay, in most cases attributed to a combination of poor construction, inadequate maintenance, and high-decay-hazard environment such as wet, poorly drained areas. This combination of decay-promoting circumstances is common throughout the United Kingdom but is by no means restricted to it. Although the actual cases of wet rot found in the survey outnumbered those of dry rot, by far the greatest proportion of damage was from dry rot; wet rot, in other words, may occur in a very limited portion of a structure, where the wood is kept constantly or periodically moist by some outside source of water, but dry rot, once it is established, can continue to progress on its own.

In the United States, the major house-rotting fungus is *Poria incrassata*, although many other species of wood-rotting fungi may be involved. *P. incrassata* does not grow so umbrageously as *Merulius lachrymans* nor does it fill up enclosed spaces with great masses of mycelium, as does *M. lachrymans*, but once it is well established in the wood it evidently can grow rapidly enough to keep up the moisture content of the wood by means of metabolic water. Not uncommonly a house or other wooden structure in the high-decay-hazard areas of the southeastern United States will be decayed almost to the point of collapse before any decay becomes evident to those occupying or using it.

Control of House Rot

As with so many other evils, prevention of house rot is much easier, cheaper, and more effective than cure after the rot already is there. Basically this means:

Use only sound, dry wood in construction. This is easy to say, but difficult for the ordinary person to do anything about, since he has no control over the quality of the material that goes into the interior framing members of a house, such as joists and studdings and rafters and roof and wall sheathing. Sometimes wood is shipped green and is partly decayed by the time it gets to the lumberyard. Sometimes it is poorly cared for in the lumberyard — piled up in the open, or inadequately protected from rain and snow. Very few of those who buy and sell lumber have any technical understanding of decay processes, and it is unlikely that they could recognize wood decay until it had progressed to the final stages; and, being like the rest of us sometimes mean, avaricious, and unscrupulous, they may deal in partly decayed materials. If the wood in question still retains sufficient strength for the use to which it is put in the structure, and if its moisture content is not then and does not later become high enough for decay to occur, a slight amount of decay may make no difference. But if that wood later becomes wet, it may decay to the point of failure much more rapidly than it would have if it had been really sound.

Second, construct the house so that there will be no leakage of water into the structure from the outside and no significant amount of condensation anywhere inside. If the house has no basement, but sits above the ground, with a crawl space underneath, the ground in this crawl space should be covered with waterproof paper or plastic, and the crawl space should be well ventilated. Regardless of how well a house is constructed, the wood in it here and there may be occasionally or regularly subjected to wetting, from condensation on toilet tanks and bowls and on cold water pipes, from rain coming in through the windows not closed before a cloudburst came, from someone running over the sink or bathtub, from backup of snow melt on the roof. Good construction must be followed by good maintenance.

Third, those parts subject to relatively high decay hazard, such as window and door frames, should be impregnated with a good wood preservative. If this is not done by the manufacturer, as it should be and often is, a liberal brush treatment of these items with a fungicide as they are installed and before they are primed for painting will serve as a good decay deterrent. A solution of pentachlorophenol in light oil is excellent for this purpose; it is available under various trade names in almost any hardware store or lumberyard. A common misconception is that paint or varnish is a preservative and that it will protect wood from decay. (See figure 8-5.) Concerning that, a Department of Agriculture bulletin (92) has this to say: ''Paint is not a preservative. In many cases it may prevent decay by protecting wood from intermittent wetting, especially if applied to ends and edges as well as exposed faces and so maintained as to allow the fewest possible cracks at joints. In some other cases, as for example when applied to wood that is not seasoned, it may favor decay by hindering further drying.'' Railroad ties and telephone poles are conspicuous examples of wood products whose service life is tremendously extended by protection from decay — but by pressure treatment with creosote or pentachlorophenol, not by a surface coating of paint. In wood used outside, as in fence posts and palings, where it is constantly exposed to moist soil or to the weather, decay prevention by means of a good preservative can mean the difference between a post rotting out in a few years and one remaining sound for several decades.

Figure 8-5. When this sign was erected, the wood was given several heavy coats of varnish to "protect" it; now, a few years later, it is decayed to the point of collapse

Once decay has become established in a building it may be necessary to remove the decayed wood, to find and eliminate the source of moisture, and perhaps to replace the decayed portion with wood treated with a preservative. In the case of dry rot some experts have recommended that the decayed wood be burned as soon as it is removed, lest it contaminate sound wood going into the construction, and even that saws and chisels used in removing the decayed wood be sterilized before being used on sound wood. This makes it sound as if dry rot were communicable by contact, like smallpox or the social diseases. As stated above, inoculum of wood-rotting fungi probably is just about universally present; if the conditions are favorable for decay to occur, it will occur, and is not likely to be prevented by washing the tools with an antiseptic.

9

Fungi Past and Present

From what, when, and how have fungi evolved, and where are they going? For thousands of years similar questions have been asked by thoughtful men about many of the higher plants and animals. One simple answer, of course, is that the fungi and all other kinds of living things were created, de novo, as described in Genesis 1:21: "And God created great whales, and every living creature that moveth, which the waters brought forth abundantly, after their kind, and every winged fowl after his kind; and God saw that it was good."

There are other views. Long before Christian times, men were puzzled about the great deposits of seashells in a thousand varieties, as well as of other fossils, high up in the mountains. To quote from Lyell (99), "Now the remains of animals, especially of aquatic species, are found almost everywhere imbedded in stratified rocks. Shells and corals are the most frequent, and with them are often associated the bones and teeth of fish, fragments of wood, impressions of leaves, and other organic substances. Fossil shells of forms such as now abound in the sea, are met with far inland, both near the surface and at all depths below it, as far as the miner can penetrate. They occur at all heights above the level of the ocean, having been observed at an elevation of from 8000 to 9000 feet in the Alps and Pyrenees, more than 13,000 feet high in the Andes, and above 15,000 feet in the Himalayas." With the development, or evolution, of Christianity and the acceptance of creation as given in Genesis, these fossils were explained in various ways, principally as follows:

1. They were a product of the Universal Deluge and Flood that was ridden out by Noah in his ark of gopherwood. Where, some doubters asked, did all the water come from, to flood the whole earth to a depth of two or three miles? That is an enormous amount of water — the earth has an area of about 200,000,000 square miles, and to flood it to a depth of, say, two miles would require, obviously, 400,000,000 cubic miles of water. Some rain! Where, the doubters also asked, did the water go to when the flood receded? Both were good questions. Where *did* the water go to?

2. The fossils were natural concretions that just crystallized out. Whether those who backed this naive theory ever examined any of these fossils firsthand is not stated. As Leonardo da Vinci remarked, concerning fossil deposits in the earth near Florence, where a canal that he had designed was being dug, some of the fossil oyster shells had the fossil remains of worms on their outside, and holes bored through the shells by predators, even as the oyster shells in the bay at the mouth of the Arno River had, and the fossil remains of fishes had teeth that were blunted and broken with wear. These crystallized out of inert salts in the soil?

Such fossils continued to bother students of nature and fathers of the church (sometimes in the same person) for many centuries, especially since both fossils and students continued to turn up in increasing numbers and variety. When the remains of dinosaurs and their relatives, and fossil ferns and cycads and palms, were found in the same rock strata in England and across the channel in France (some of the fossil dinosaurs and pterodactyls with the fossilized remains of prehistoric prey in their fossilized stomachs), it became a bit difficult to use the Noah's ark explanation. Dinosaurs and pterodactyls on the ark? Lyell, about whom more will be said shortly, describes fossil forests, mainly of cycads, the precursors of conifers, in the deposits on the Isle of Portland, off the coast of England, and mentions the "celebrated lithographic stone" of Bavaria in which a fellow geologist had collected 237 species of fossils, among them 7 distinct species of flying lizards or pterodactyls, 6 of saurians, 3 of tortoises, 60 of fish, 56 of crustacea (lobsters, shrimps, crabs, barnacles, and their kindred), and 26 of insects. By then — this was in the 1830s — the idea of a gradual development of different kinds of living things over a relatively long stretch of time was at least gaining ground among some biologists, and was shortly to gain much more ground much more rapidly with the publication of Charles Darwin's *Origin of Species* in 1859. It may be interesting to go into the background of this a bit because, while it does not deal directly with the fungi, some acquaintance with it will enable readers to see how the evidence was gathered and why we have so few solid data concerning the course of evolution of fungi.

The idea of organic evolution did not spring forth fully formed all of a sudden from Darwin's powerful mind, nor was it a product especially of his observations of the similarities and differences among related species of finches and turtles on neighboring islands of the Galapagos when he visited them in 1835, as described in his delightful *Voyage of the Beagle* (94). Earlier in the book he gives accounts of the remains of prehistoric animals that he had unearthed a couple of years before in Argentina and that he recognized as the ancestors of some of the present-day animals peculiar to South America. He wrote, "This wonderful relationship in the same continent between the dead and the living, will, I do not doubt, hereafter throw more light on the appearance of organic beings on our earth, and their disappearance from it, than any other class of facts." When he left England in December 1831, at the age of 22, he was still a fundamentalist, supposedly, and accepted the doctrine of special creation as stated somewhat generally in Genesis but in much greater detail by various commentators, and he probably also accepted the official church dictum on the age of the earth — it was formed in 4004 B.C. at 9:00 A.M. on September 21. Now, two years later, he recognizes that some of the kinds of animals presently in South America have descended from other and in some ways quite different kinds. Formerly, he says, the land must have swarmed with great monsters; there were fossil anteaters, armadilloes, tapirs, peccaries, guanacos, opossums, and others, some of them much larger than their present-day descendants. In describing the geography of Patagonia and nearby regions he mentions that the mind is stupified in thinking over the length of time that had to elapse for the changes to occur. When he found seashells high in the mountains he recognized them as seashells that had been formed when the land was sea bottom, the land subsequently having been uplifted and made into mountains.

What had happened to convert this orthodox and proper young man (at Cambridge he had studied for the ministry and it was thought that he would become a country parson) into an ardent evolutionist? For one thing, probably, his own observation and study of plants, animals, and geography and geology in South America — as I stated earlier, Charles Darwin was a bright young man. He also was a keen observer, and combined this with a very good brain to absorb and digest what he saw. He himself gives a good deal of credit for his concept of evolution to Charles Lyell's book *Principles of Geology*, which someone had given him to take along on the voyage, with the admonition that although it was interesting, Darwin was not to believe what was in it, or to accept Lyell's unorthodox views. Good advice, perhaps, from one fundamentalist to another, but futile, because no one could read the evidence from geology and paleontology presented by Lyell and fail to grasp the fact that over a tremendous period of time the earth and the living things on it have been constantly undergoing changes. Especially was this true when, as in Darwin's case, he could see firsthand the evidences of the very changes that Lyell so clearly and convincingly described. From this it seems that Darwin was a professing evolutionist long before he wrote *The Origin of Species*.

He was by no means the first one — the thought that there had been a gradual development of existing species from preexisting ones had been blowing around in the biological community for well over a century, and enough was spoken and written on the subject to warrant the writing of several books on the history of the concept of evolution up to the time of Charles Darwin (93, 98, 100). These different books do not all agree on the importance of the contributions of different men to the development of the theory of evolution — one giving Buffon, in France, credit for having anticipated Darwin, and praising him highly, and another more or less pooh-poohing Buffon's contributions. Most of Buffon's ideas on the subject are scattered here and there through a set of twenty-odd volumes of an encyclopedia of biology which he wrote over a period of several decades, and it would be a wonder if his views did not change some in the course of years, and if, at one time, he did not make statements directly contradictory to those he had made much earlier. But it is true that he at different times made so many statements on different sides of the case that it is hard to tell where he really stood. And, besides, many of his statements were not backed up by any solid evidence. The major credit should go to the man who not only enunciates the principle but sets forth enough facts to substantiate it, rather than to the man who makes all sorts of statements and then, when the principle is established, pulls out of his files the one statement he made 20 years before that supports his claim to priority, waves it in the air, and shouts, "See — that's what I said years ago!" In any case, Buffon was a prolific and popular writer and, if nothing else, he got across the idea that, given time enough, organic beings were capable of almost unlimited change.

Charles Darwin's grandfather, Erasmus Darwin, was a keen observer of nature and, in various publications shortly before and after 1800, suggested that change in form and function of living things is possible over a period of generations and also presented some astute ideas on how the changes could come about. Many of these observations and ideas are in a two-volume work, *Zoonamia*, and others are embodied in two long poems and in the footnotes that accompany them, a somewhat novel way of presenting new concepts in science. Jean Baptiste Lamarck, a Frenchman, a few years later published in much less circumspect form similar ideas on how evolution might occur — so similar, in fact, that some writers have accused him of having cribbed them from Erasmus Darwin without acknowledgment. This is possible, but unlikely, since Erasmus Darwin's ideas on evolution were pretty well hidden in his long-winded discussions of other things, and who expects to find revelations in the footnotes to interminably long poems? Charles Darwin, in his *Origin of Species*, merely says that it is remarkable how closely his grandfather anticipated Lamarck. Some have contended that Charles Darwin did not give Lamarck or others of his predecessors sufficient credit for their contributions to his theory of evolution, but this seems unfair. Ideas evolve like almost everything else; besides, Darwin's great individual accomplishment was not in *enunciating* the theory of evolution but in suggesting the mechanism of it, in accumulating, over a period of nearly 30 years, a tremendous quantity of data to support the theory, and in subjecting the mass of data to rational analysis that enabled him to construct a carefully thought-out scheme of nature.

Of course he was indebted to many predecessors, as he himself recognized. Much of the groundwork necessary for the development of the theory of evolution was done by geologists, not by botanists or biologists as such, although a hundred and fifty to two hundred years ago, when these concepts were developing, there were not such sharp distinctions between different disciplines as have arisen since. Some who were primarily geologists in those days were very knowledgeable in paleontology and in botany and zoology and biology in general. That these men should be the ones to lay the groundwork for the later evolutionists was only natural: they were the ones who came in direct contact with the evidence of life of the past.

Darwin had other precursors, of whom two deserve some mention, mainly for the preposterousness of their claims. One of them, by the name of Wells, was an expatriate American who, in 1813, wrote an essay entitled "On Dew," which he delivered before the august and elite Royal Society, and which supposedly contained all the essentials of Darwin's theory of natural selection. So what? It is one thing to propound a theory more or less out of the blue, as Wells did, and another thing to laboriously and intelligently collect evidence over a period of 25 years and then develop a theory to explain the facts that have been accumulated. Poets and other seers and soothsayers long ago foretold travel by air, and even interplanetary flight. They didn't build a plane or a spaceship, though. Any moderately imaginative person when stimulated by hunger and perhaps a modicum of strong water can envision great new developments, but until he gets the time machine actually functioning in interplanetary overdrive he is just dealing in fantasy. Suppose that some day someone actually does come up with a spaceship equipped with interplanetary overdrive — he himself, the man who designed it and built it, would deserve the credit for it, not the man who, way back when, suggested that it could be done. So with Wells and Darwin. Wells did not receive much recognition or credit from Darwin, but he didn't deserve much, either.

Another precursor, mostly self-recognized, was one Patrick Mathew, a crabby Scot who in 1831, 28 years before the *Origin of Species* was published, in a book entitled, of all things, *Naval Architecture and Arboriculture*, outlined the principles of natural selection as propounded by Darwin. After the *Origin of Species* was published, Mathew became so wrought up over the fact that he was not given credit for the idea that he had cards printed and distributed stating that he, not Darwin, deserved recognition for having originated the theory. Mathew by the way, did not mention Wells, who had preceded him by 18 years — he, like most of us, was concerned mainly with *his* not being awarded his suitable measure of glory, and not with awarding his predecessors *their* proper measure of glory.

Both Wells and Mathew had the same amount of time available to them that Darwin had in which to develop a body of evidence to support their nebulous theory, yet they did nothing. Why not? If they were so good, what held them back? In my opinion they got about as much credit as they deserved.

As Eiseley remarks (95), the astronomers discovered infinite space in the seventeenth century. Up to near that time the earth was considered to be the center of the universe, surrounded by a solid firmament ("Let there be plenty of firmament!" shouts the Lord in *Green Pastures*), and the sun and stars and planets circled around it. Galileo in the 1500s was imprisoned for stating that the earth moved around the sun and was forced to recant that horrid heresy. With Newton's discovery of universal gravitation the solar system became a self-sustaining, self-regulating, perpetual, and eternal engine that no longer needed the personal intervention of God to keep it functioning, and early in the 1700s Halley, for whom the comet is named, showed that our planet earth was adrift in a great star swirl. The idea of infinite time developed more slowly. James Hutton, a geologist, in the late 1700s measured the amount of silt in running streams and, from that simple procedure, calculated the rate of denudation of the land and the rate of accumulation of sedimentary deposits, and postulated the infinitely slow subsidence and uplifting of land mass through an immense period of time.

William "Strata" Smith was a surveyor, engineer, and field geologist who knew the lay of the land in England better than anyone else of his time (1769–1839). He observed that different strata bore the remains of different kinds of plants and animals, and also realized that, if the strata were in their original sequence, the lower ones would be the oldest and would contain the more ancient fossils, and the upper ones would be younger and would contain the more recent fossils. In progressing from the lower and older to the upper and newer strata in a given deposit it was possible to follow the course of development of different kinds of plants and animals, the rise and fall of populations, and, in short, the course of evolution. Also, by means of fossils characteristic of a given deposit, a given stratum, it was possible to recognize that stratum wherever it cropped out. Application of this principle showed that some deposits had been laid down over vast areas. Smith opened up the past to exploration by paleontologists, and they have been exploring it assiduously ever since. For those interested in learning more about what this exploration has led to, the book *Life of the Past* (86) is recommended.

Charles Lyell (1797–1875) was primarily a geologist, but also a competent biologist. He presented a tremendous amount of very detailed data on the individual species and plant and animal communities of the distant past. He recognized that competition among species was a factor in survival, and he granted that some species became extinct — the evidence of the rocks was unmistakable on that point. But when it came to the origin of new species from older ones, through gradual change and the survival of the fittest, he hesitated and equivocated. It has been said that Darwin's *Origin of Species* could have been written from Lyell's *Principles of Geology*, but that is only good hindsight, and it disregards the fact that one of the key points needed for the formulation of the theory of organic evolution by Darwin was missing from Lyell's work — namely, the mode of origin of new species.

Basically, Darwin's theory consists of the following principles: (1) All species are varying in all characteristics at all times and everywhere. (2) All species produce more offspring than can survive. (3) In the struggle for existence, the favored races survive. He did not, at that time, have the benefit of much information on genetics or the laws of inheritance, or on cytology and the physical basis of inheritance and variation, but he did a fabulous job in the *Origin of Species*, and it took the world by storm.

Why, although some ideas concerning evolution were published before then, and probably had been spread by word of mouth for hundreds of years, did it take so long for them to jell into a firmly established theory? For one thing, there was relatively little study of nature, just for the sake of trying to find out more about it, until well into the 1600s. For many centuries, questions about natural phenomena were answered by consulting the sacred works or the approved authorities, which meant the Bible and commentaries on it, and the works of Aristotle. Until the invention of movable type in the 1400s, books were written by hand, on parchment, and only by "clerks" or other members of the religious communities who could read and write. New ideas, good or bad, could not be readily promulgated and spread around and hashed over and subjected to the fire of criticism. Up until almost 1800 in many countries just about anything of a scholarly nature had to be subjected to the approval of the church, and in questions about the nature of the solar system and the structure of the earth and the descent of man, the Bible was the supreme authority. To many it still is. Recently two branches of a Lutheran synod have split over the question of literal interpretation of the Bible, and no longer ago than 1973, three young biologists wrote a letter to *Phytopath News*, a sort of comings-and-goings publication of the American Phytopathological Society, to the effect that if the theory of organic evolution was to be taught, then the biblical account should be given equal time. Their biology may be sound, although one wonders how much of Darwin or Lyell or of the comparative anatomists and embryologists and paleontologists they have come in contact with, but their scholarship concerning the Bible probably does not amount to much. Andrew D. White, an extremely able scholar, was president of and professor of history at Cornell University for some time before 1900. He also was a very staunch pillar of the Episcopal church, and one of his deep interests was the history of religions, particularly the Christian religion, including the Bible. His *History of the Warfare of Science with Theology in Christendom* (he distinguishes sharply between theology and religion) is superb, and certainly should be required reading for anyone who pretends to any scholarship concerning the Holy Writ. In the latter part of Volume II of the work he cites by name some of the greatest religious scholars of the times, both Catholic and

Protestant, to the effect that much of the Old Testament, and especially Genesis, evolved from much more ancient Chaldean, Assyrian, Egyptian, and Asian religious writings and myths. Anyone who chooses to question this should at least take the trouble to try to find out for himself the basis of their arguments.

So much for that. Now, a bit more about the time required for evolution to have occurred by means of gradual changes such as we see in the populations of animals and plants around us. Just about when it seemed that the proponents of evolution had been given all the time they needed, the rug was rudely jerked from under them. Erasmus Darwin had postulated ''millions of ages'' for the course of evolution, without specifying what an age was. The geologists from Hutton on and, later, the paleontologists, all pretty well agreed that a vast amount of time was required for the slow wearing away and building up of the earth and for the rise and fall of populations of plants and animals long extinct. A question often asked, by the way, is what change, slow and gradual or sudden and cataclysmic, led to the extinction of the dinosaurs that for a long time must have been a dominant form of animal life, in myriad shapes and sizes. No mystery is involved at all: what happened to them and to most of their associates is simply what has happened to the great majority of species of plants and animals throughout the course of evolution. A few fellow travelers of the dinosaurs still are with us — alligators and turtles among the animals, and araucarias, cycads, and tree ferns among the plants, changed some but still recognizable. And cockroaches, for example, got where they wanted to go ages ago and have not changed much since. But the normal fate of most species is eventual extinction — either in time they evolve into something different, in their slow adjustment to the changes in the web of life of which they are a part, or they do not adjust, in which case they expire. Many more species have become extinct than have survived, an obvious biological truism that at times seems difficult to grasp. So, if there now are somewhere around two million species of living things on the earth, which seems a fair estimate, then throughout the course of evolution there must have been quite a few million more. And if all this occurred by changes so gradual that even the greatest of them is scarcely detectable from one generation to the next, a good deal of time is required. By the time the *Origin of Species* was published, hundreds of millions of years did not seem out of line, and even a billion years probably would have been accepted as reasonable.

Then came a shock. In the 1860s Lord Kelvin, leading physicist of the time and by some said to be the greatest physicist of the nineteenth century, determined, by measuring "residual heat," that the earth could not be more than ten million years old, at most. A physicist, of course, could actually measure something (in this case, temperature) with an instrument (in this case, a thermometer), make his calculations by straightforward arithmetic (figures don't lie), and come up with hard facts. The biologists and geologists and paleontologists dealt in nebulous assumptions and surmises and estimates. They were guessing. Kelvin *knew*. He had made his determination by precise measurements and precise calculations, and that was it. If his conclusion torpedoed the ship of evolution so painstakingly constructed by the biologists and geologists and paleontologists, that was just too bad, because he was RIGHT. And arrogantly right, in the bargain. Never, evidently, did it occur to him to question his methods or his conclusions, which was too bad, because he was totally wrong. Present-day physicists, using measurement of radioactivity in the rocks, put the age of the earth at around five and a half to six billion years, and the age of the oldest known remains of any living thing (presumably a primative alga) at about three billion years. That seems long enough to have permitted the forms of life we see today, and those of which there are fossil records, to have evolved by gradual change. However, if there are today about two million species of living things and if, throughout the course of evolution, there have been many more — say, conservatively, ten million — then that means that a new species arose on the average every 300 years. Yet in the time that man has been observing species fairly carefully — say, over a period of several thousand years — no one has actually recorded the appearance of a new species of any kind. This does not mean that no new species have arisen, since not all living things have been neatly tagged and catalogued, and many new species may have arisen in that time without our being aware of it; certainly new varieties and races of a multitude of different forms of life, from viruses and bacteria to horses and cows and, probably, man, have appeared in that time, and new variants continue to appear. Also, whether new species are appearing depends partly on our definition of a species.

Another thing to be considered is that, at present, just about all ecological niches are filled, and a new type of any species, from microbes to man, might have a relatively tough time making it. In earlier times, say two to three billion years ago, this was not so. We do not know what the environment was then, or even where the present continents were then, but in a mostly unoccupied world new forms of life must have had more *Lebensraum* and much less competition than a new form has now. Of course we are assuming that the present-day physicists cannot possibly be in error in *their* estimates of the age of the earth.

Since Darwin's time evidence about the fact of evolution has been accumulated from so many different sources that only fanatics could refuse to accept the conclusions — that is, the present species on the earth, including fungi and man, developed from preexisting species through descent with modification, and the major selecting force has been competition in the struggle for existence, as Darwin postulated. Comparative anatomy, including vestigial organs, embryology, cytology, genetics (the maize plant, about the origin of which there has been a good deal of discussion over the years, has actually been bred "backwards" to disclose the plants from which it evolved), biochemistry, paleontology — all point to the same conclusion. The embryos of man, monkeys, fish, and fowl at an early stage of development are indistinguishable from one another. Biochemically the molds are very similar to man — their proteins are composed of the same amino acids, and they metabolize their foods in the same way. The geological or paleontological record has established the lines of descent of quite a few kinds of plants and animals with assurance, and dating by radioactive materials has established the extreme age of some of the ancestral lines. For a modern summary of organic evolution, see the books by Hotton (98) and Moody (100).

So where did the fungi come from? For many years, from about 1870 on, most biologists in all fields felt obligated to work out a scheme of evolution for those particular groups of organisms with which they were occupied. There were not then, and still are not, many remains of fungi from ancient times. From the first development of living things, whenever and wherever that occurred, decay and dissolution must have been about as essential as they are today — one process does not go on very long without the other. According to this argument, fungi probably are about as ancient as any other form of life similarly organized, that is, with nuclei. There must, of course, have been a long period of evolution leading up to the development of something so precisely machined and organized as nuclei.

For quite a few decades, fungi were derived by the mycologists from the algae — the Phycomycetes from the green algae, the Ascomycetes from the red algae, and the Basidiomycetes from the Ascomycetes. The algae developed in and inhabited the shallow seas and, later, freshwater ponds and lakes, as they still do, and somewhere along the line they developed chlorophyll.

A modern theory, by the way, and one with a good deal of evidence to support it, is that the present organisms with organized nuclei and other organelles within the cell probably developed from combinations of different organisms long, long ago and that what started out as various sorts of partnerships between distinct kinds of organisms progressed gradually to a balanced state in which the *combination* of two or more organisms survived and operated as one. The chloroplasts in the cells of green plants, for example, and also the mitochondria, contain their own special and peculiar type of DNA and protein-manufacturing system, different from the DNA and protein-manufacturing system of the nucleus of the cell. Both of these probably once were "guests" or parasites or partners within the primitive cell. This is by no means as farfetched as it may sound. A one-celled protozoan that lives as a symbiont in the gut of a termite (and which digests the wood that the termite lives on, and so provides food for its host) is in turn host to several different kinds of external and internal symbionts that are so closely bound up with it that they constitute part of the cell. Another single-celled protozoan, this one a free-living *Paramecium*, has within it numerous individuals of a single-celled green alga, and under favorable conditions the photosynthetic processes of these symbionts furnish the *Paramecium* with its food. The green alga can grow independently outside of the *Paramecium*, but the *Paramecium* cannot live without the green alga within it. So organisms such as algae, in this scheme, probably evolved from a combination of several more primitive things, one of which contributed chlorophyll to permit use of sunlight in manufacturing food. They in turn gave rise to the higher plants.

Probably no modern microbiologist believes that the algae gave rise to the fungi. Algae and fungi look somewhat alike — or, more accurately, some fungi are structurally similar to some algae, although it is difficult to see much resemblance between a red seaweed and a mushroom. The old idea that the algae lost chlorophyll and thereby became fungi is simple to the point of ingenuousness; such a conversion would require an entirely new set of enzymes, a different way of life, a total reorganization of vital physiological and biochemical processes. Albinos are common in many higher plants, but do not, just by loss of chlorophyll, become fungi — they simply die off from lack of nourishment, although in the laboratory they can be kept alive for a time by special feeding. In the enjoyable *Life of the Past*, which gives so good an account of the course of evolution of animals, fungi are disposed of in a single sentence to the effect that they are derived from algae by loss of chlorophyll, for which the author probably should be forgiven because he is primarily a student of animals, although he could have questioned this old view. Now, the fungi are in a kingdom of their own, entirely separate from that of the Protista, in which the algae, of whatever shape, size, and color, are included, and separate from the kingdom Plantae, which includes the higher plants. When did they evolve? We don't know. Many of them are extremely specialized, as you must know by now, but we have no knowledge of how these specialists developed, or from what. Many mycologists speculate about the origin and development of this or that group of fungi. An outgoing president of the Mycological Society of America, in fact, is almost expected to do this. It allows him to display his erudition and profound scholarship, and may even be interesting in a way, but nobody takes it too seriously. The course of evolution of any given group of organisms is determined mainly by a combination of approaches — fossil remains in rocks, comparative morphology of gross and detailed parts, embryological development, chemical affinities, and so on. As mentioned above, there are few ancient fossil remains of fungi; fungi much like some of those in existence today have been described on and within the roots of Cycads of carboniferous times, 250 to 280 million years ago, but there just are not many fossil fungi. Also fungi do not have much morphology: they are structurally simple. They have no embryology at all, unless one wishes to consider the germ tube of a germinating spore to be an embryo. Chemical studies help to show us that man and molds are related to one another, but other than that give us little or no information about the course of evolution of the fungi. It seems reasonable to assume that some of the precursors of the present fungi evolved along with the precursors of other things now living, and so they probably have been around for a considerable time. From their varied mode of life and their ability to shift and roll with the blows of fate, it also seems likely that they will be around for quite some time longer, probably a good deal longer than man.

So although we do not have much information on the course of evolution of fungi, we have accumulated a great deal of information on their way of life at present. Man probably has eliminated some species of fungi, since he has eliminated a number of species of animals and it is likely that at least some of those animals had fungi peculiar to themselves on their outsides or insides; but none of the environmentalists, so far as I know, have put any of the fungi on the list of endangered species. The major function of fungi in the biological world is to convert the remains of most plants, and of some animals, into their constituent elements to be used again by the generations that follow. Some of them have expanded this beneficent life style and attack and digest living plants and animals, and many of them live on raw and manufactured products of all kinds. In general, though, the fungi live in relative harmony with their environment and with their fellow living beings, which is more than can be said of man. For a long time Western man, at least, has followed the biblical injunction to use and dominate and exploit the earth. "And God said, Let us make man in our image, after our likeness; and let them have dominion over the fish of the sea, and over the fowl of the air, and over the cattle, and over all the earth, and over every creeping thing that creepeth upon the earth" (Genesis I:26). This has been called the "Christian Ethic," but some of the more advanced non-Christians seem to pursue the same plan just about as effectively as do the most devout and orthodox Christians. In the olden days this policy of total domination and exploitation of all living things by man, for his own immediate benefit, may have seemed reasonable, just, and right, and to many it still seems so, although even the least perceptive must realize that we cannot continue this irresponsible exploitation forever. We have pretty well done away with quite a few fowl of the air, and are fairly well along on the fishes of the sea. We humans hope that as a species man can continue to survive for at least some tens of thousands of years yet (a hope that, if they could hope, probably would not be shared by many other species on the earth) but whether or not we do, and, if we do, how well, will depend to some extent upon how well we integrate ourselves with our fellow travelers, including some of the fungi.

This is not to suggest that an understanding of how fungi live and grow, and a knowledge of what they do and how they fit into the over-all scheme of nature, on the part of people in general will in itself make for the good life. But if all people had a much deeper and broader knowledge of an appreciation of other sorts of living things, including fungi, they might develop more respect for nature and a feeling of being just a part of the intricate web of life, a partner and participant in the struggle, instead of a brutal master with the divine injunction to mutilate, use up, and destroy anything in their path. Of course we have to swing the balance of nature in our favor by controlling noxious weeds and pests, but we can at the same time recognize the right of other things to live, even if they do not on the surface appear to be made in God's image, as we so obviously are — but some of us, no doubt, more so than others.

Maybe man has long passed the stage where he can fit harmoniously into the ecology; he may be so committed to a life style of exploitation and consumption that he cannot change. It is ironic that the have-not or developing countries are demanding a greater share of the goods and goodies of the world just when the goods and goodies are running out and, at the same time, the populations of those countries that can least support or endure large populations are expanding at an ever-increasing rate.

So where do the fungi fit into this? Probably most importantly as agents of destruction of plants, animals, and all sorts of raw and manufactured materials. Any reduction of such losses is a net gain for the people of the earth. Even in some of the developed countries, losses of stored grains and grain products, losses which could easily be prevented, are very high; they are disregarded simply because of indifference, lack of interest, or official inertia. In some regions in some years plant diseases take a tremendous toll; we must learn how to anticipate and avoid these destructive attacks — not just by fungicidal sprays but by all means at our disposal including, it is to be hoped, ecologically acceptable ones. Have we explored all possible approaches in plant-disease control? After all, varieties of plants were being bred for specific disease resistance for about 50 years before someone thought to develop, by the same approach, plants with generalized resistance to many different pathogens, an approach now being given a big push.

Fungus diseases of animals, as mentioned in the earlier chapter dealing with that subject, are by no means the killers that some of the parasitic bacteria and viruses are, but they still are important enough to deserve more attention than they have received. Probably much more important, from an economic standpoint, are the mycotoxins — and with these potentially and actually damaging compounds produced by fungi we have had only a bit more than a decade of experience; in our knowledge of them and of their effects upon us and our domestic animals we are about where bacteriology was a century ago.

Fungi also are used for food, either as themselves, in the way of wild and cultivated mushrooms, or as ripening agents, as in the conversion of various starchy plant materials into "vegetable cheeses" (97). These have been developed in and continue to be used almost only in southeast Asia. Perhaps they are worth looking into as another possible food source. One aspect of the relation of fungi to nutrition is the possible production by the fungi of some sorts of growth-stimulating compounds that could be used as feed additives. As was mentioned in *The Molds and Man* (42), insects such as leaf-cutting ants, some castes of some kinds of termites, and ambrosia beetles live a full and complete life with no other food than the peculiar fungi that they cultivate — and, like most other living things, those insects need proteins, carbohydrates, vitamins, and minerals, and must get them from the fungi that they consume. Maybe this would be worth looking into by a team competent in mycology, fungus physiology, and nutrition.

Among the drugs, the antibiotic penicillin and the various alkaloids of ergot are the only major ones got from fungi, although as was noted in chapter 4 a number of other fungi, and especially some of those in the genus *Fusarium*, produce a whole gamut of compounds that are very active physiologically in some kinds of animals. This field also could use some intensive investigation by teams of experts in the required disciplines.

Obviously, mycology is entering into our lives in a fairly big way. For a long time mycology was mainly the describing of new species and the working out of systems of classification, but that no longer is true. The nomenclaturists will continue to coin new names, in the fond delusion that thereby they are Pushing Back the Barriers, and the classificationists will continue to construct new and ever more intricate and involved systems of classfication, but mycology has in the last generation or so become much more that that, and it seems reasonable to predict that the mycological horizons will in the future continue to expand rapidly into new, now unknown, exciting, and interesting fields.

References

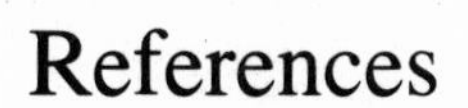

References

Chapter 1. Poisonous and Hallucinogenic Mushrooms

1. Ainsworth, G. C., and G. R. Bisby. *A Dictionary of the Fungi*. 4th ed. Kew, Surrey, England: Commonwealth Mycological Institute, 1954.
2. Allegro, J. M. *The Sacred Mushroom and the Cross*. Garden City, N.Y.: Doubleday, 1970.
3. Barron, F., M. E. Jarvik, and S. Bunnel, Jr. "The Hallucinogenic Drugs," *Scientific American*, 210:29–37 (1964).
4. Christensen, C. M. *Common Edible Mushrooms*. Minneapolis: University of Minnesota Press, 1943; 1972 reprinting.
5. Christensen, C. M. *Edible Wild Mushrooms*. Extension Bulletin 357, Agricultural Extension Service, University of Minnesota. St. Paul, 1968.
6. Fischer, O. E. "Mushroom Poisoning," pp. 825–864 in Kauffman's *Agaricaceae of Michigan*, cited in reference 10.
7. Floersheim, G. L. "Curative Potencies against A-Amanitin Poisoning by Cytochrome C," *Science*, 177:808–809(1972).
8. Güssow, H. T., and W. S. Odell. *Mushrooms and Toadstools*. Ottawa, Canada: Division of Botany, Dominion Experimental Farms, 1928.
9. Harner, M. J., ed. *Hallucinogens and Shamanism*. New York: Oxford University Press, 1973.
10. Kauffman, C. H. *The Agaricaceae of Michigan*. Michigan Geological and Biological Survey, Publication 26, Biological Series 5, 1918. Reprinted in 1971 by Dover Publications, New York.
11. Krieger, L. C. C. *A Popular Guide to the Higher Fungi (Mushrooms) of New York State*. New York State Museum Handbook No. 11. Albany: University of the State of New York, 1935.
12. Miller, O. K. *Mushrooms of North America*. New York: E. P. Dutton, 1972.
13. Pilat, A., and Otto Usak. *Mushrooms*. Amsterdam: H. W. Bijl, 1954.
14. Ramsbottom, J. *Mushrooms and Toadstools*. London: Collins, 1953.
15. Sanford, J. H. "Japan's 'Laughing Mushrooms,'" *Economic Botany*, 26:174–181 (1972).

16. Schultes, R. E. "Teonanacatl: The Narcotic Mushroom of the Aztecs," *American Anthropologist*, n.s., 42:429–443 (1940).
17. Seaver, F. J. *The North American Cup-Fungi (Inoperculates)*. Published by the author, 1951
18. Simons, D. M. "The Mushroom Toxins," *Delaware Medical Journal*, 43:177–187 (1971).
19. Smith, A. H. *The Mushroom Hunter's Field Guide*. Ann Arbor: University of Michigan Press, 1963.
20. Wasson, R. G. "The Hallucinogenic Mushrooms of Mexico: An Adventure in Ethnomycological Exploration," *New York Academy of Sciences Transactions*, 21:325–339 (1958–1959).
21. Wasson, R. G., and V. P. Wasson. *Mushrooms, Russia and History*. 2 vols. New York: Pantheon Books, 1957.

Chapter 2. Ergot and Ergotism

22. Barger, G. *Ergot and Ergotism*. London: Gurney and Jackson, 1931.
23. Bovè, F. J. *The Story of Ergot*. Basel, Switzerland: S. Karger AG, 1970.
24. Dinusson, W. E., C. N. Haugse, and R. D. Knutson. "Ergot in Rations for Fattening Cattle," *Farm Research* (North Dakota State University), 20–21 (November–December 1971).
25. Gröger, D. "Ergot," chapter 12 in *Microbial Toxins*, vol. 8: *Fungal Toxins*, ed. S. J. Ajl *et al.* New York: Academic Press, 1972.
26. Seaman, W. L. *Ergot of Grains and Grasses*. Publication 1488, Research Station, Ottawa, Ontario, Canada, 1971.
27. U.S. Department of Agriculture, Agricultural Marketing Service, Grain Division. *Official Grain Standards of the United States*. Revised. Washington, D.C.: Government Printing Office, February 1970.

Chapter 3. Mycotoxins and Mycotoxicoses: Aflatoxin

28. Brewington, C. R., and J. L. Weirach. "Survey of Commercial Milk Samples for Aflatoxin M," *Journal of Dairy Science*, 53:1509–1510 (1970).
29. Burnside, J. E., W. L. Sippel, J. Forgacs, W. T. Carll, M. B. Atwood, and E. R. Doll. "A Disease of Swine and Cattle Caused by Eating Moldy Corn: II. Experimental Production with Pure Cultures of Molds," *American Journal of Veterinary Research*, 18:817–824 (1957).
30. Christensen, C. M. "Pure Spices — How Pure?" *American Society for Microbiology News*, 38:165–169 (1972).
31. Christensen, C. M., H. A. Fanse, G. H. Nelson, Fern Bates, and C. J. Mirocha. "Microflora of Black and Red Pepper," *Applied Microbiology*, 15:622–626 (1967).
32. Goldblatt, L., ed. *Aflatoxin*. New York: Academic Press, 1969.
33. Keyl, A. C., A. N. Booth, M. S. Masri, M. R. Gumbmann, and W. E. Gagne. "Chronic Effects of Aflatoxin in Farm Animal Feeding Studies," pp. 72–75 in *Toxic Micro-organisms: Mycotoxins — Botulism*, ed. M. Herzberg. UJNR Joint Panels on Toxic Micro-organisms and the U.S. Department of the Interior. Washington, D.C.: Government Printing Office, 1970.
34. Kraybill, H. F., and R. E. Shapiro. "Implications of Fungal Toxicity to Human Health," chapter 15 in Goldblatt's *Aflatoxin*, cited in reference 32.
35. Moreno-Martinez, E., and C. M. Christensen. "Fungus Flora of Black and White Pepper (*Piper nigrum* L.)," *Revista Latina-Americana de Microbiologia*, 15:19–22 (1973).

36. Raper, K. B., and Dorothy I. Fennell. *The Genus Aspergillus*. Baltimore: Williams and Wilkins, 1965.
37. Shank, R. C. "Dietary Aflatoxin Loads and the Incidence of Human Hepatocellular Carcinoma in Thailand," pp. 245–262 in *Symposium on Mycotoxins in Human Health*, ed. I. F. H. Purchase. New York: Macmillan, 1971.
38. Shotwell, Odette L., C. W. Hesseltine, H. R. Burmeister, W. F. Kwolek, G. M. Shannon, and H. H. Hall. "Survey of Cereal Grains and Soybeans for the Presence of Aflatoxin: 1. Wheat, Grain Sorghum and Oats," *Cereal Chemistry*, 46:446–454 (1969).
39. Shotwell, Odette L., C. W. Hesseltine, H. R. Burmeister, W. F. Kwolek, G. M. Shannon, and H. H. Hall. "Survey of Cereal Grains and Soybeans for the Presence of Aflatoxin: 2. Corn and Soybeans," *Cereal Chemistry*, 46:454–463 (1969).
40. Uraguchi, K. "Mycotoxic Origin of Cardiac Beriberi," *Journal of Stored Products Research*, 5:227–236 (1969).
41. Van Walbeek, W., P. M. Scott, and F. S. Thatcher. "Mycotoxins from Food-Borne Fungi," *Canadian Journal of Microbiology*, 14:131–137 (1968).

Chapter 4. Mycotoxins and Mycotoxicoses: Other *Aspergillus* Species, *Penicillium*, and *Fusarium*

See also references 35, 36, 40.

41a. Carlton, W., J. Tuite, and P. Mislivec. "Pathology of the Toxicosis Produced in Mice by Corn Cultures of *Penicillium viridicatum*," pp. 94–106 in *Proceedings of the 1st U.S.-Japan Conference on Toxic Micro-Organisms*, ed. M. Herzberg. Washington, D.C.: Government Printing Office, 1970.
42. Christensen, C. M. *The Molds and Man*. 3rd. ed. Minneapolis: University of Minnesota Press, 1965.
43. Christensen, C. M., C. J. Mirocha, G. H. Nelson, and J. F. Quast. "Effect on Young Swine of Consumption of Rations Containing Corn Invaded by *Fusarium roseum*," *Applied Microbiology*, 23:202 (1972).
44. Joffe, A. Z. "Toxin Production by Cereal Fungi Causing Toxic Alimentary Aleukia in Man," pp. 77–85 in *Mycotoxins in Foodstuffs*, ed. G. N. Wogan. Cambridge, Mass.: MIT Press, 1965.
45. Krogh, P., and E. Hasselager. "Studies on Fungal Nephrotoxicity," pp. 198–214 in *Yearbook of the Royal Veterinary and Agricultural College*. Copenhagen, 1968.
46. Kulik, M. M. *A Compilation of Descriptions of New Penicillium Species*. U.S. Department of Agriculture, Agriculture Research Service, Agriculture Handbook No. 351. Washington, D.C.: Government Printing Office, 1968.
47. Nelson, G. H., C. M. Christensen, and C. J. Mirocha. "A Veterinarian Looks at Moldy Corn," pp. 86–91 in *Proceedings of the 20th Annual Hybrid Corn Industry Research Conference*. 1965.
48. Raper, K. B., and C. Thom. *A Manual of Penicillia*. Baltimore: Williams and Wilkins, 1949. Facsimile edition, New York: Hafner Publishing Co., 1968.
49. Shank, R. C., G. N. Wogan, J. B. Gibson, and A. Nondasuta. "Dietary Aflatoxins and Human Liver Cancer: II. Aflatoxins in Market Foods and Foodstuffs of Thailand and Hong Kong," *Food Cosmetology and Toxicology*, 10:61–69 (1972).
50. Smalley, E. B., W. F. O. Marasas, and Susan Daugherty. *Development of Aspergillus Flavus and Aflatoxins in Harvested Forage*. University of Wisconsin College of Agricultural and Life Sciences Research Bulletin R2412, 1972.

51. Smalley, E. B., W. F. O. Marasas, F. M. Strong, J. R. Bamber, R. E. Nichols, and N. R. Kosuri. "Mycotoxicoses Associated with Moldy Corn," pp. 163–173 in *Proceedings of the 1st U.S.–Japan Conference on Toxic Micro-Organisms*, ed. M. Herzberg. Washington, D.C.: Government Printing Office, 1970.
52. Snyder, W. C., and H. N. Hansen. "The Species Concept in *Fusarium*," *American Journal of Botany*, 27:64–67 (1940).
53. Snyder, W. C., and H. N. Hansen. "The Species Concept in *Fusarium* with Reference to Section Martiella," *American Journal of Botany*, 28:738–742 (1941).
54. Snyder, W. C., and H. N. Hansen. "The Species Concept in *Fusarium* with Reference to Discolor and Other Sections," *American Journal of Botany*, 32:657–666 (1945).
55. Theron, P. J., R. J. van der Merwe, N. Liebenberg, H. J. B. Joubert, and W. Nel. "Acute Liver Injury in Ducklings and Rats as a Result of Ochratoxin Poisoning," *Journal of Pathology and Bacteriology*, 91:521–529 (1966).
56. Toussoun, T. A., and P. E. Nelson. *A Pictorial Guide to the Identification of Fusarium Species*. University Park: Pennsylvania State University Press, 1968.

Chapter 5. Airborne Fungus Spores, Plant Disease, and Respiratory Allergy

57. DuBuy, H. G., and Mary D. Lackey. *A Comparative Study of Sampling Devices for Air-Borne Micro-Organisms*. U.S. Public Health Service, Public Health Reports, Supplement No. 184, 1945.
58. Durham, O. C. "An Unusual Shower of Fungus Spores," *Journal of the American Medical Association*, 111:24–25 (1938).
59. Durham, O. C. "Results of Mold Surveys by Slide Examinations," in Feinberg's *Allergy in Practice*, cited in reference 60.
60. Feinberg, S. M. *Allergy in Practice*. 2nd ed. Chicago: The Yearbook Publishers, 1946.
61. Glick, P. A. *The Distribution of Insects, Spiders, and Mites in the Air*. U.S. Department of Agriculture Technical Bulletin 673, 1939.
62. Gregory, P. H. *The Microbiology of the Atmosphere*. 2nd ed. New York: John Wiley and Sons, 1973.
63. Gregory, P. H., J. M. Hirst, and F. T. Last. "Concentrations of Basidiospores of the Dry Rot Fungus (Merulius Lachrymans) in the Air of Buildings," *Acta Allergologica*, 6:168–174 (1953).
64. Hirst, J. M. "Changes in Atmospheric Spore Content: Diurnal Periodicity and the Effects of Weather," *Transactions of the British Mycological Society*, 36:375–393 (1953).
65. Ingold, C. T. *Spore Liberation*. Oxford: Clarendon Press, 1965.
66. Ingold, C. T. *Fungus Spores: Their Liberation and Dispersal*. Oxford: Clarendon Press, 1971.
67. Ingold, C. T., and V. J. Cox. "Periodicity of Spore Discharge in *Daldinia*," *Annals of Botany*, 29:201–209 (1955).
68. Pady, S. M., and L. Kapica. "Fungi in Air over the Atlantic Ocean," *Mycologia*, 47:34–50 (1955).
69. Stakman, E. C. "Progress and Problems in Plant Pathology," *Annals of Applied Biology*, 42:22–33 (1955).
70. Weston, W. H. "Production and Dispersal of Conidia in the Philippine Sclerosporas of Maize," *Journal of Agricultural Research*, 23:239–278 (1923).

Chapter 6. Fungus Predators and Parasites of Nematodes and Insects

71. Bassi, Agostino. *Del Mal del Segno*, trans. P. J. Yarrow. Phytopathological Classics No. 10. Baltimore: American Phytopathological Society, 1958.
72. Drechsler, C. "Some Hyphomycetes That Prey on Free-Living Terricolous Nematodes," *Mycologia*, 29:447–552 (1937).
73. Duddington, C. L. *The Friendly Fungi*. London: Faber and Faber, 1957.
74. Petch, T. "Fungi Parasitic on Scale Insects," *Transactions of the British Mycological Society*, 7:18–40 (1920).
75. Shanor, L. "Some Observations and Comments on the Laboulbeniales," *Mycologia*, 47:1–12 (1955).
76. Steinhaus, E. A. "Microbial Control — the Emergence of an Idea," *Hilgardia*, 26:107–160 (1956).

Chapter 7. Fungi Pathogenic in Man and Animals

77. Ainsworth, G. C. *Medical Mycology: An Introduction to Its Problems*. New York: Pitman, 1952.
78. Ainsworth, G. C. *Ainsworth and Bisby's Dictionary of the Fungi*. 5th ed. Kew, Surrey, England: Commonwealth Mycological Institute, 1961.
79. Austwick, P. K. C. "Pathogenicity," chapter 7 in *The Genus Aspergillus*, by K. B. Raper and Dorothy I. Fennell. Baltimore: Williams and Wilkins, 1965.
80. Conant, N. F., D. S. Martin, D. T. Smith, R. D. Baker, and J. L. Callaway. *Manual of Clinical Mycology*. Philadelphia: W. B. Saunders Co., 1945.
81. Dawson, Christine O. "Ringworm in Animals," *Review of Medical and Veterinary Mycology*, 6:223–233 (1968).
82. DeMonbreun, W. A. "The Cultivation and Cultural Characteristics of Darling's Histoplasma Capsulatum," *American Journal of Tropical Medicine*, 14:93–125 (1934).
83. Fiese, M. J. *Coccidioidomycosis*. Springfield, Ill.: C. C. Thomas, 1958.
84. Georg, Lucille K. *Animal Ringworm in Public Health: Diagnosis and Nature*. U.S. Public Health Service Publication No. 727. Washington, D.C.: Government Printing Office, 1960.
85. Jungerman, P. F., and R. M. Schwartzman. *Veterinary Medical Mycology*. Philadelphia: Lea and Febiger, 1972.
86. Simpson, G. G. *Life of the Past: An Introduction to Paleontology*. New Haven, Conn.: Yale University Press, 1954.
87. Skinner, C. E., C. W. Emmons, and H. M. Tsuchiya. *Henrici's Molds, Yeasts, and Actinomycetes*. 2nd ed. New York: John Wiley and Sons, 1947.
88. Sweany, H. C., ed. *Histoplasmosis*. Springfield, Ill.: C. C. Thomas, 1960.

88a. Staib, F. "Aspergillus fumigatus in der Ausatmungsluft eines Arztes," *Deutsche Medizinische Wochenschrift*, 99:1804–1807 (1974).

88b. Helm, E. B., W. Lunkenbein, and W. Stille. "Aspergillus-fumigatus-Infektionen in der Inneren Medizin," *Deutsche Medizinische Wochenschrift*, 99:1807–1810 (1974).

Chapter 8. Decay of Wood in Trees and Buildings

89. Boyce, J. S. *Forest Pathology*. 2nd ed. New York: McGraw-Hill, 1948.
90. Findlay, W. P. K. *Dry Rot and Other Timber Troubles*. London: Hutchinson, 1953.

91. Hickin, N. E. *The Dry Rot Problem*. London: Hutchinson, 1963.
92. U.S. Department of Agriculture. *Decay and Termite Damage in Houses*. Farmers' Bulletin 1993, 1948; revised 1951.

Chapter 9. Fungi Past and Present

See also references 42, 86

93. Carlile, M. C., and J. J. Skehel, eds. *Evolution in the Microbiological World.* Cambridge: At the University Press, 1974.
94. Darwin, Charles R. *The Voyage of the Beagle*. Everyman's Library. New York: E. P. Dutton, 1967 reprinting.
95. Eiseley, Loren. *Darwin's Century: Evolution and the Men Who Discovered It*. Garden City, N.Y.: Doubleday, 1958.
96. Gray, W. D. *The Relation of Fungi to Human Affairs*. New York: Henry Holt, 1959.
97. Gray, W. D. *The Use of Fungi as Food and in Food Processing*. Cleveland, Ohio: CRC Press, 1970.
98. Hotton, Nicholas, III. *The Evidence of Evolution*. New York: American Heritage, 1968.
99. Lyell, Charles. *Elements of Geology*. 2nd American ed. Pittsburgh: James Kay, Jun. & Brother, 1848. (This is a condensed version of Lyell's *Principles of Geology*.)
100. Moody, P. A. *Introduction to Evolution*. 3rd ed. New York: Harper and Row, 1970.
101. Osborn, Henry F. *From the Greeks to Darwin*. 2nd ed. New York: Charles Scribner's Sons, 1929.
102. White, Andrew D. *A History of the Warfare of Science with Theology in Christendom*. 2 vols. New York: D. Appleton, 1898. (Also available in Modern Library edition.)

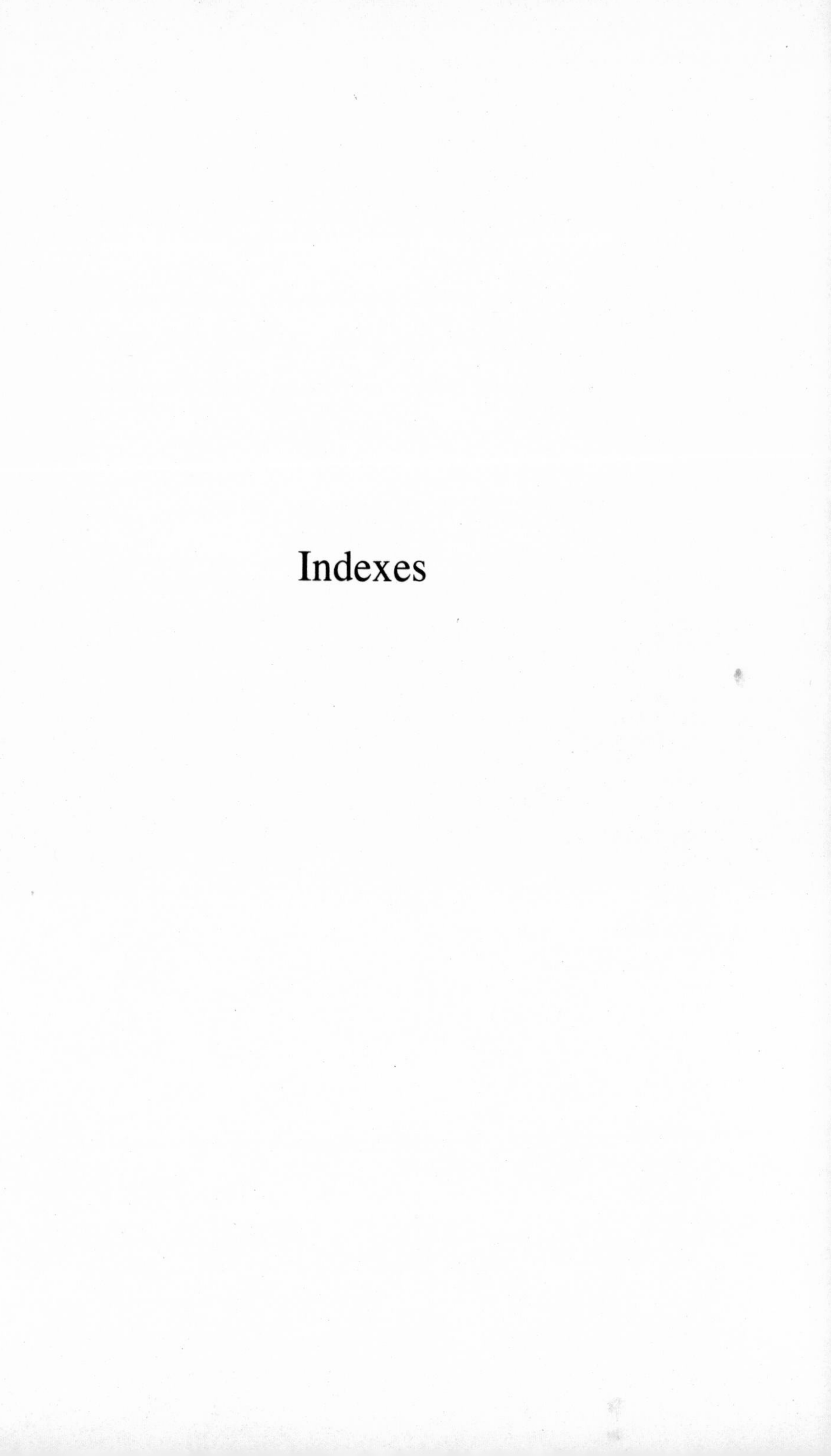

Indexes

Subject Index

Index of Names, Including Authors Cited

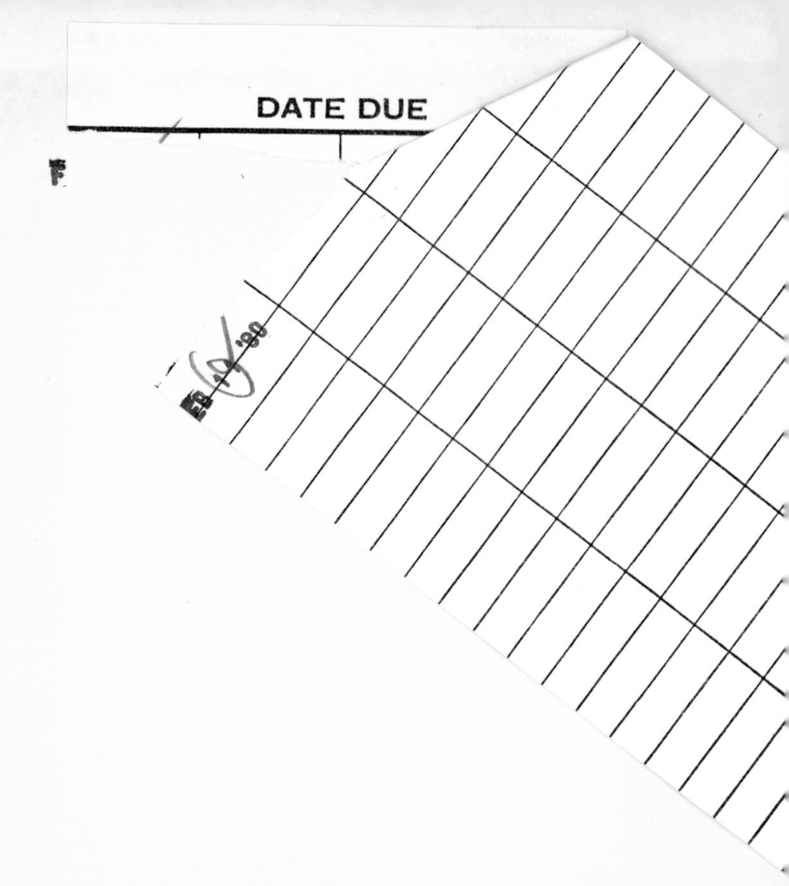

Christensen, Clyde Martin, 1905-

Molds, mushrooms, and mycotoxins / by Clyde M. Christensen. — Minneapolis : University of Minnesota Press, [1975]

264 p. : ill. ; 23 cm.

Includes bibliographies and indexes.

ISBN 0-8166-0743-5

1. Fungi, Pathogenic. I. Title.

309506 QR245.C48 1975 589'.2'0465 74-21808
MARC

Library of Congress 75